AF595011

Advances in Chemical Analysis Procedures (Part I)

Advances in Chemical Analysis Procedures (Part I): Extraction and Instrument Configuration

Special Issue Editors

Marcello Locatelli
Angela Tartaglia
Dora Melucci
Abuzar Kabir
Halil Ibrahim Ulusoy
Victoria Samanidou

MDPI • Basel • Beijing • Wuhan • Barcelona • Belgrade

Special Issue Editors

Marcello Locatelli
Department of Pharmacy Build B, level 2; University "G. d'Annunzio" of Chieti-Pescara
Italy

Angela Tartaglia
Department of Pharmacy, University "G. d'Annunzio" of Chieti-Pescara
Italy

Dora Melucci
Department of Chemistry "Giacomo Ciamician", University of Bologna
Italy

Abuzar Kabir
International Forensic Research Institute, Department of Chemistry and Biochemistry, Florida International University
USA

Halil Ibrahim Ulusoy
Analytical Chemistry, Faculty of Pharmacy, Cumhuriyet University
Turkey

Victoria Samanidou
Laboratory of Analytical Chemistry, School of Chemistry, Aristotle University of Thessaloniki
Greece

Editorial Office
MDPI
St. Alban-Anlage 66
4052 Basel, Switzerland

This is a reprint of articles from the Special Issue published online in the open access journal *Molecules* (ISSN 1420-3049) in 2019 (available at: https://www.mdpi.com/journal/molecules/special_issues/chemical_analysis_extraction_instrument_configuration).

For citation purposes, cite each article independently as indicated on the article page online and as indicated below:

LastName, A.A.; LastName, B.B.; LastName, C.C. Article Title. *Journal Name* **Year**, *Article Number*, Page Range.

ISBN 978-3-03936-577-7 (Hbk)
ISBN 978-3-03936-578-4 (PDF)

Contents

About the Special Issue Editors

Marcello Locatelli earned his degree in Chemistry from the University of Bologna, Department of Chemistry "G. Ciamician" with a thesis on the "Development of an analytical methodology for the analysis and identification of protein adducts by mass spectrometry". He earned his Ph.D. from the University of Bologna, Department of Chemistry "G. Ciamician" with a thesis on the "Combined analytical methods of mass spectrometry for the study of impurities in drugs and metabolites of biomolecules". He is also an Associate Professor in Analytical Chemistry at the University "G. d'Annunzio" of Chieti-Pescara, Department of Pharmacy.

His research activity is devoted to the development and validation of chromatographic methods for the qualitative and quantitative determination of biologically active molecules in complex matrices from human and animal (whole blood, serum, plasma, bile, tissues, feces, and urine), cosmetics, foods, and environmental. This activity comprises the study of all processes related to pre-analytical stages such as sampling, extraction and purification, separation, enrichment, and even the application of conventional and hyphenated analytical methods for accurate, sensitive, and selective determination of biologically active molecules. Recently, particular attention has been focused on innovative (micro)extraction procedures like MEPS, FPSE, MIP, DLLME, and SULLE. These procedures have been applied to different compounds, from synthetic and natural origin (glucosamine, 5-amino-salicylic acid, natural or synthetic bile acids, anti-inflammatory, drugs association and fluoroquinolones, secondary metabolites from natural sources, heavy metals). Predictive models and chemometric are also tested in the development of the method, both for the optimization of extraction protocols and final data processing. Particular attention is focused on new instrument configurations for quantitative analysis in a complex matrix.

Prof. Locatelli's contribution to the field includes more than 157 manuscripts, 116 congress communications, 1 patent subject to approval, 13 book chapters and 3 books, and he has served as a guest editor for 13 Special Issues with demonstrable scientific impact (based on Scopus (8th of June 2020), h-index 33, 148 papers, 2778 citations). Besides, he is a reviewer of the following international journals: Analytica Chimica Acta, Current Bioactive Compounds, Journal of Chromatography A, Talanta, Trends in Analytical Chemistry (more than 100 International peer-reviewed Journals). He is a referee for the MIUR Institution for National Projects (SIR), and included in the register REPRISE (Register of Expert Peer Reviewers for Italian Scientific Evaluation) in the "Basic Research" section. He is a referee for VQR 2011-2014. He is also a referee for other universities of proposals through competitive tenders for the allocation of University funds for the activation of research grants. He is an Editorial Board member of the journal Molecules section "Analytical Chemistry", Current Analytical Chemistry, Separations, Current Bioactive Compounds, American Journal of Modern Chromatography, Journal of Selcuk University Science Faculty, Review in Separation Sciences, Cumhuriyet Science Journal. He is an Associate Editor of the magazine Frontiers in Pharmacology section "Ethnopharmacology", Review Editor of the journal Frontiers in Oncology section "Pharmacology for anti-cancer drugs", and Review Editor of the journal Frontiers in Medical Technology section "Nano-Based Drug Delivery". He is a member of the Scientific Committee of the journal Scienze e Ricerche, published by the Italian Book Association.

ADDITIONAL TITLES

Member of the Italian Chemical Society (SCI, card number 13779);

Member of the American Chemical Society (ACS, card number 30617260),

Member of the Italian Society of Toxicology (Sitox),
Member of the Italian Society of Phytochemistry (SIF).

Angela Tartaglia obtained her master's degree in Pharmacy in March 2018 from the University "G. d'Annunzio" of Chieti (Italy) with a research thesis on "FPSE-HPLC-DAD method for the quantification of anticancer drugs in human whole blood, plasma, and urine". From March 2018 to June 2018 she joined the Aristotle University of Thessaloniki—School of Chemistry—Laboratory of Analytical Chemistry for additional work on microextraction procedures and method validation focused on food products. Since November 2018 she has been a Ph.D. candidate at University "G. d'Annunzio", Chieti (Italy).

Her Ph.D. research activity is focused on the optimization of new protocols for sample preparation (MEPS, FPSE, MIP) and the validation of new analytical methods for the qualitative and quantitative determination of small drugs in complex matrices, mostly biological fluids (plasma, urine, whole blood, saliva). Another of her research interests concerns natural products and the study of biologically active products from plants with beneficial properties for human health, mostly phenolic compounds due to their positive impact on health by reducing the risk of cardiovascular.

Dora Melucci obtained her Master of Science in Chemistry at the University of Bologna, with a thesis entitled "Determination of polypeptides by HPLC". She obtained a Master in Chemical Methodologies for Control and Analysis at the University of Bologna, with a thesis entitled "The gravitational field-flow fractionation technique (GrFFF). Fractionation and absolute quantitative analysis of particulate in dispersion". Finally, she obtained her Ph.D. in Chemical Sciences at the University of Ferrara, with a thesis entitled "Characterization of polymers by means of thermal field-flow fractionation (ThFFF) using decalin as a solvent".

She was an Appointed Researcher and Assistant Professor in 1999 (branch Analytical Chemistry) at the Department of Chemistry "Giacomo Ciamician", School of Sciences, University of Bologna.

Her research interests have been focused on the following: Separation science: FFF of macromolecules in solution and dispersed micro-particles. Standardless and absolute analysis in ETA-AAS and HPLC. ThFFF of industrial polymers in collaboration with industry. Flow-FFF and GrFFF of real samples (starch, yeast, metal nanoparticles for bio-static materials), miniaturization of the separation tool, cell-sorting, hyphenation of FFF with chemiluminescence, development of multi-analyte competitive immuno-enzymatic methods using dispersed nano-particles and micro-particles. In the framework of these subjects, she was the Coordinator of the Local Unity of Bologna in a National project (PRIN) entitled "Microfluidic separation of nanosystems".

Since 2005, she has been conducting an autonomous research line, with basic keyword: chemometrics. The topic concerns the application of chemometrics to the development and validation of innovative analytical methodologies, with a focus on direct and non-altering methods of analysis. The fields of application are food, environment, pharmaceutics, forensics, biotechnology, and cultural heritage. The analytical techniques employed are AAS, NIR, Voltammetry, Raman, GC, LC-MS, and FT-IR, XRD. As of May 2020, her work has produced 76 articles, 20 contributions in books, and over 100 communications in national and international meetings, with h-index = 17.

Between 2002–2020, she has had teaching responsibility in Analytical Chemistry, Laboratory of Analytical Chemistry, Analytical Chemistry and Law, Principles of Quality and Safety (Bachelor in Chemistry); Chemometrics (Master of Science in Chemistry); Chemometrics for Forensic Analysis

(Master in Forensic Chemical and Chemical-Toxicological Analysis).
She also holds the following academic roles: Local Coordinator for the University of Bologna of the National Ministerial Project for high-school student guidance (Scientific Degrees Plan), and the Department Delegate for university student tutoring.

Abuzar Kabir is a Research Assistant Professor at the Department of Chemistry and Biochemistry, Florida International University, Miami, Florida, USA. His research focusses on the synthesis, characterization, and applications of novel sol-gel derived advanced material systems in the form of chromatographic stationary phases, surface coatings of high-efficiency microextraction sorbents, nanoparticles, microporous and mesoporous functionalized sorbents, molecularly imprinted polymers for analyzing trace and ultra-trace level concentration of polar, medium polar, nonpolar, ionic analytes, heavy metals, and organometallic pollutants from complex sample matrices. His inventions, fabric phase sorptive extraction (FPSE), dynamic fabric phase sorptive extraction (DFPSE), capsule phase microextraction (CPME), molecular imprinting technology, superpolar sorbents, in-vial microextraction (IVME), sol-gel based reversed-phase LC stationary phases and SPE sorbents, organic polymeric LC stationary phases and SPE sorbents, synthesis of mesoporous silica and its application in reversed-phase LC stationary phases and SPE sorbents have drawn global attention. He has developed and formulated numerous high-efficiency sol-gel hybrid inorganic–organic sorbents based on Silicon, Titanium, Zirconium, Tantalum, Germanium chemistries. Dr. Kabir has authored 18 patents, 10 book chapters, 70 journal articles, and 125 conference papers. His recent inventions, Biofluid Sampler (BFS) and Universal Biofluid Sampler (UBFS) are capable of handling whole blood (5–1000 μL) without any sample pre-treatment for chromatographic separation and analysis. These technologies will likely change the current practices of blood analysis in the near future.

Halil Ibrahim Ulusoy received his Master's Degree in 2007 and his Doctorate Degree in 2012 on the field of Analytical Chemistry. He is a Full Professor of Analytical Chemistry at the Cumhuriyet University, Faculty of Pharmacy (Sivas/TURKEY) since 2015. He is a member of the Turkish Chemical Society. His research interests are in the development of new analytical methodologies for trace organic and inorganic species in the food samples, pharmaceutical samples, and biological matrices. His research activity is devoted to the development and validation of chromatographic and spectroscopic methods for trace determination of biologically active molecules and elements in complex matrices cosmetics, foods, and environmental samples. Recently, his academic studies are particularly focused on easy applicable and reliable determination drug active ingredients such as antibiotics, antidepressants, pesticides, and vitamins.
More than 65 manuscripts, 96 congress communications, 5 chapters in scientific books, and guest editor for 4 special issues attested scientific activity (based on Scopus (9th of June 2020), h-index 18, 55 papers, 653 citations). He's is Editorial Board member of the journal *Current Analytical Chemistry, Journal of Quality Assurance and Pharma Analysis (IJQAPA), Pharmaceutica Analytica Acta, Asian Journal of Medicinal and Analytical Chemistry*. He is one of the chief editors of the *Cumhuriyet Science Journal*.

Victoria Samanidou was born on the 11th of January 1963, in Thessaloniki, Greece. She obtained her Bachelor of Science degree in Chemistry, in 1985, from the Chemistry Department of Aristotle University of Thessaloniki, Greece. From 20-7-86 to 25-8-86 she was at the Institute of Ecological Chemistry, in GSF, Attaching/Freising, Germany, for additional work on her Ph.D. Thesis, as well as research work on Photochemistry and the study of photodecomposition products of chlorophenols by HPLC-Diode array and GC-MS. From 15-7-87 to 4-9-87 she was at the Institute of Ecological Chemistry, in GSF, Neuherberg-Munich, Germany, for additional work on her Ph.D. Thesis, as well as research work on carbamate analysis by HPLC and GC-MS. From 1-7-88 to 30-9-88 she was at the Institute of Ecological Chemistry, in GSF, Neuherberg-Munich, Germany, for additional work on her Ph.D. Thesis, as well as research work on controlled release of pesticides by HPLC and GC-MS.
In 1990, she obtained a doctorate (Ph.D.) in Chemistry from the Department of Chemistry of the Aristotle University of Thessaloniki. The topic of her thesis was "Distribution and mobilization of heavy metals in waters and sediments from rivers in Northern Greece". In the same year, Dr. Samanidou joined the Laboratory of Analytical Chemistry, in the Department of Chemistry of the Aristotle University of Thessaloniki, as a Technical Assistant. Nine years later she was appointed as a Lecturer in the Laboratory of Analytical Chemistry in the Department of Chemistry of the Aristotle University of Thessaloniki. In 2007, she joined the Institute of Analytical Chemistry and Radiochemistry in Graz Technical University, for four months, developing methods by LC-MS/MS.
Since 2015, Dr. Samanidou is a Full Professor in the Laboratory of Analytical Chemistry in the Department of Chemistry of the Aristotle University of Thessaloniki, Greece, where she currently serves as Director of the Laboratory.
Dr. Samanidou has authored and co-authored more than 170 original research articles in peer-reviewed journals and 45 reviews and 50 chapters in scientific books, with H-index 36 (Scopus June 2020, http://orcid.org/0000-0002-8493-1106, Scopus Author ID 7003896015) and ca 3500 citations. She has supervised four Ph.D. Theses, 24 postgraduate Diploma Theses, 2 postdoc researchers, and more than 15 undergraduate Diploma Theses. She has served as a Member of 10 advisory Ph.D. committees, 21 examination Ph.D. committees, and 32 examination committees for postgraduate Diploma Theses. She is a member of the editorial board of more than 10 scientific journals and she has reviewed ca 500 manuscripts in more than 100 scientific journals. She was also a guest editor in more than 10 Special Issues in scientific journals. She has served as an Academic Editor for *Separations* MDPI, as Regional editor in *Current Analytical Chemistry* and as Editor-in-Chief of *Pharmaceutica Analytica Acta.*
Her research interests include the following: 1. Development and validation of analytical methods for the determination of inorganic and organic substances using chromatographic techniques. 2. Development and optimization of methodologies for sample preparation of various samples, e.g., food and biological fluids. 3. Study of new chromatographic materials used in separation and sample preparation (polymeric sorbents, monoliths, carbon nanotubes, fused core particles, etc.) compared to conventional materials.
She was also a Member of the organizing and scientific committee in 20 scientific conferences.
Since December 2015, Dr. Samanidou has been President of the Steering Committee of the Division of Central and Western Macedonia of the Greek Chemists' Association. In November 2018, she was re-elected to serve in the same leading position for 3 more years.
A milestone in her career was when she was included in the top 50 power list of women in Analytical Science in 2016, as proposed by Texere Publishers.

Preface to "Advances in Chemical Analysis Procedures (Part I): Extraction and Instrument Configuration"

Innovative analytical protocols are needed during all processes involved in chemical analysis, to establish the identities, properties, pureness, bioactivities, quantities, and other main features. Furthermore, modern analytical methods provide accurate information about therapeutic dosage and/or pharmacokinetics about new chemical entities (NCE). During the last few years, new techniques for chemical analysis have been developed that allow separation scientists to get all the relevant information quickly and accurately even when the analysis is carried out in complex matrices (biological fluids, environmental samples, or food matrices). This Special Issue will highlight and describe the latest innovations in instruments and techniques developed for the extraction, analysis, and characterization of drugs, bioactive compounds, and pollutants in samples that, until recently, were difficult to analyze. Greater attention will be paid to recent developments in sample preparation and extraction techniques, with a focus on modern detection techniques, starting with one of the stalwarts of analytical chemistry laboratories, high-performance liquid chromatography (HPLC), up to more innovative instrumentation such as mass spectrometry, mostly used in biological sciences for detection and quantization of protein.

Marcello Locatelli, Angela Tartaglia, Dora Melucci, Abuzar Kabir, Halil Ibrahim Ulusoy, Victoria Samanidou

Special Issue Editors

 *molecules* 

Article

Fast Detection of 10 Cannabinoids by RP-HPLC-UV Method in *Cannabis sativa* L.

Mara Mandrioli [1], Matilde Tura [1], Stefano Scotti [2] and Tullia Gallina Toschi [1,*]

[1] Department of Agricultural and Food Sciences, Alma Mater Studiorum-University of Bologna, Viale Fanin 40, 40127 Bologna, Italy; mara.mandrioli@unibo.it (M.M.); matilde.tura2@unibo.it (M.T.)

[2] Shimadzu Italia, Via G. B. Cassinis 7, 20139 Milano, Italy; sscotti@shimadzu.it

* Correspondence: tullia.gallinatoschi@unibo.it; Tel.: +39-051-209-6010

Academic Editor: Marcello Locatelli
Received: 10 May 2019; Accepted: 31 May 2019; Published: 4 June 2019

Abstract: Cannabis has regained much attention as a result of updated legislation authorizing many different uses and can be classified on the basis of the content of tetrahydrocannabinol (THC), a psychotropic substance for which there are legal limitations in many countries. For this purpose, accurate qualitative and quantitative determination is essential. The relationship between THC and cannabidiol (CBD) is also significant as the latter substance is endowed with many specific and non-psychoactive proprieties. For these reasons, it becomes increasingly important and urgent to utilize fast, easy, validated, and harmonized procedures for determination of cannabinoids. The procedure described herein allows rapid determination of 10 cannabinoids from the inflorescences of *Cannabis sativa* L. by extraction with organic solvents. Separation and subsequent detection are by RP-HPLC-UV. Quantification is performed by an external standard method through the construction of calibration curves using pure standard chromatographic reference compounds. The main cannabinoids dosed (g/100 g) in actual samples were cannabidiolic acid (CBDA), CBD, and Δ9-THC (Sample L11 CBDA 0.88 ± 0.04, CBD 0.48 ± 0.02, Δ9-THC 0.06 ± 0.00; Sample L5 CBDA 0.93 ± 0.06, CBD 0.45 ± 0.03, Δ9-THC 0.06 ± 0.00). The present validated RP-HPLC-UV method allows determination of the main cannabinoids in *Cannabis sativa* L. inflorescences and appropriate legal classification as hemp or drug-type.

Keywords: cannabinoids; *Cannabis sativa* L.; HPLC; validation

1. Introduction

Cannabis is classified into the family of Cannabaceae and initially encompassed three main species: *Cannabis sativa*, *Cannabis indica*, and *Cannabis ruderalis* [1]. Nowadays, Cannabis has only one species due to continuous crossbreeding of the three species to generate hybrids. In fact, all plants are categorized as belonging to *Cannabis sativa* and classified into chemotypes based on the concentration of the main cannabinoids. Depending on the THCA/CBDA ratio, some chemotypes have been distinguished. In particular, chemotype I or "drug-plants" have a TCHA/CBDA ratio >1.0, plants that exhibit an intermediate ratio are classified as chemotype II, chemotype III or "fiber-plants" have a THCA/CBDA ratio <1.0, plants that contain cannabigerolic acid (CBGA) as the main cannabinoid are classified as chemotype IV, and plants that contain almost no cannabinoids are classified as chemotype V [2–5].

Recently, in Italy the interest in *Cannabis sativa* L. has increased mainly due to the latest legislation (Legge n. 242 del 2 dicembre 2016) [6]. As a consequence, there is a request to develop cost-effective and easy-to-use quantitative and qualitative methods for analysis of cannabinoids.

The Italian regulatory framework has classified two types of *Cannabis sativa* L. depending on the content of Δ9-THC. In particular, fiber-type plants of *Cannabis sativa* L., also called "hemp", are characterized by a low content of Δ9-THC (<0.2% *w/w*). If the content of Δ9-THC is >0.6% *w/w*, it is considered as drug-type, also called "therapeutic" or "marijuana".

Industrial hemp is used in several sectors, such as in the pharmaceutical, cosmetic, food, and textile industries, as well as in energy production and building. In general, fiber-type plants are less used in the pharmaceutical field, where drug-type plants are more often employed [5]. However, there is also an increased interest in hemp varieties containing non-psychoactive compounds. In fact, the European Union has approved 69 varieties of *Cannabis sativa* L. for commercial use [7].

Hemp has a complex chemical composition that includes terpenoids, sugars, alkaloids, stilbenoids, quinones, and the characteristic compounds of this plant, namely cannabinoids. *Cannabis sativa* L. has several chemotypes, each of which is characterized by a different qualitative and quantitative chemical profile [5]. The cannabinoids, terpenes, and phenolic compounds in hemp are formed through secondary metabolism [3,8]. The term "cannabinoid" indicates terpenophenols derived from *Cannabis*. More than 90 cannabinoids are known, and some are derived from breakdown reactions [8]. Gaoni and Mechoulam [9] were the first to define cannabinoids "as a group of C_{21} compounds typical of and present in *Cannabis sativa*, their carboxylic acids, analogs, and transformation products". Currently, cannabinoids have been classified according to their chemical structure, mainly seven types of cannabigerol (CBG); five types of cannabichromene (CBC); seven types of cannabidiol (CBD); the main psychoactive cannabinoid Δ9-tetrahydrocannabinol (Δ9-THC) in nine different forms including its acid precursor (Δ9-tetrahydrocannabinolic acid, Δ9-THCA); Δ8-tetrahydrocannabinol (Δ8-THC), which is a more stable isomer of Δ9-THC but 20% less active; three types of cannabicyclol (CBL); five different forms of cannabielsoin (CBE); seven types of Cannabinol (CBN), which is the oxidation artifact of Δ9-THC; cannabitriol (CBT); cannabivarin (CBDV); and tetrahydrocannabivarin (THCV) [10,11]. THC, CBD, CBG, CBN, and CBC are not biosynthesized in *Cannabis sativa*, and the plant produces the carboxylic acid forms of these cannabinoids (THCA, CBDA, CBGA, CBNA, and CBCA). Cannabinoid acids undergo a chemical decarboxylation reaction triggered by different factors, mainly temperature. This decarboxylation reaction leads to the formation of the respective neutral cannabinoids (THC, CBD, CBG, CBN, and CBC) [12,13].

There are several methods to quantify cannabinoids [14–21], some of which require expensive mass spectrometry detectors [22–25]. Furthermore, there is a great deal of uncertainty around the use of gas chromatography (GC) for the titration of cannabinoids due to the high temperature of the injector and detector that can lead to the decarboxylation of cannabinoid acids if not derivatized correctly [26]. Moreover, recent studies have reported that cannabinoid acid decarboxylation is only partial, and as result the actual value is underestimated. An HPLC system allows for determination of the actual cannabinoid composition, both neutral and acid forms, without the necessity of the derivatization step [13].

It is necessary, in addition to honed methods, to develop new procedures with a view to discriminate different *Cannabis* varieties in order to identify and titrate cannabinoids in a simple way. These methods should ideally be fast, easy, robust, and cost-efficient as they can be used not only by research laboratories but also by small companies with a view on quality control.

This study focuses on the development, validation, and step-by-step explanation of a rapid and simple HPLC-UV method for identification and quantification of the main cannabinoids in hemp inflorescences that can be easily reproduced and applied. The method described is focused on the quantification of CBD but can also be applied to check the levels of THC.

2. Results and Discussion

2.1. Method Development

The aim of this work was to develop a new analytical method for determination of the main cannabinoids in hemp samples. In fact, the method described below can be used as a routine quality control procedure and can be applied by the pharmaceutical industry, small laboratories, or even small pharmacies.

A crucial aspect for accurate identification and quantification of analytes is optimization of separation conditions, and therefore various preliminary tests were carried out (e.g., mobile phase, detection wavelength). Different mobile phases were tested, and trials were performed with different compositions and gradient elution to optimize the separation of all 10 target compounds considered (File S2). The greatest difficulty was that of separating CBD and THCV, which in many cases co-eluted. It was also difficult to separate the isomers Δ9-THC and Δ8-THC. The best resolution of cannabinoids was obtained using a chromatographic column and, as an eluent mixture, water with 0.085% phosphoric acid and acetonitrile with 0.085% phosphoric acid.

The quantification of cannabinoids was made at 220 nm after testing different wavelengths (File S2). This wavelength represents the best compromise for all the cannabinoids considered and was selected to detect and integrate all compounds of interest within the dedicated concentration range. As far as chromatographic analysis is concerned, before using the instrument, the system was conditioned for 20 min by fluxing the eluent mixture in the instrument under the same conditions as the method, and then a chromatographic run was performed by injecting 5 μL of acetonitrile to verify that the chromatographic system was adequately cleaned. Simultaneously with the analysis of the sample, standard solutions were injected at different concentrations for the construction of calibration curves and to evaluate the separation and identification of each compound. The identification of cannabinoids was performed by comparing their retention times with those obtained by the injection of pure standards and by an enhancing procedure. Figure 1 shows a chromatogram of a standard mixture of cannabinoids and Figure 2 shows a chromatogram of a sample of hemp.

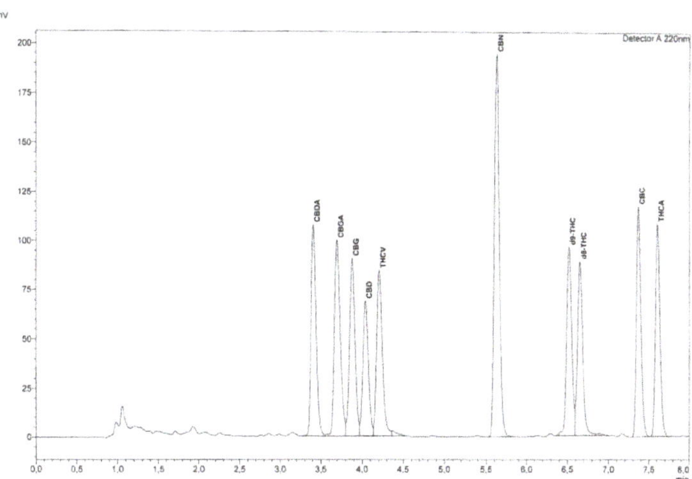

Figure 1. Chromatographic trace of a standard cannabinoid mixture analyzed by RP-HPLC-UV equipped with reverse phase C18 column.

Cannabinoids in different varieties of *Cannabis sativa* L. can be present in very different concentrations. In order to obtain good chromatographic separation and correct quantification, it may be necessary to dilute or concentrate the extract, performing two different injections. For example, in the case of high levels of CBDA or CBD it will be necessary to dilute the extract. For THC, it is often found at low concentration in hemp inflorescences, so it may be necessary to concentrate the extract before injection. In our case, 2 mL of filtered extract was dried using a weak nitrogen flow, and the dry extract was recovered in 500 μL of acetonitrile.

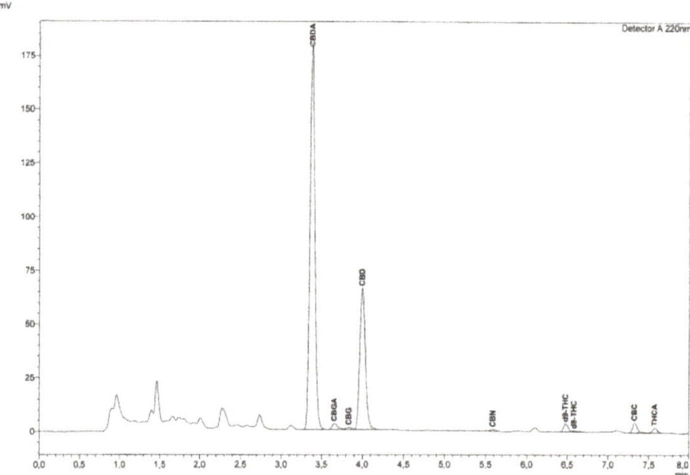

Figure 2. Chromatographic trace of *Cannabis sativa* L. inflorescence extract analyzed by RP-HPLC-UV equipped with a reverse phase C18 column.

2.2. Validation

2.2.1. Precision

The precision of the method was measured by the expression of repeatability (*r*) and reproducibility (*R*). Precision was expressed through coefficient of variation (CV%).

2.2.2. Repeatability, R

Table 1 shows data on the intraday and interday repeatability, evaluated as reported in Section 3.6, which demonstrates very high repeatability. In fact, the relative standard deviation (RSD) varied from 2.59 to 5.65 for intraday repeatability and from 2.83 to 5.05 for interday repeatability. In both cases, the highest RSD was found for CBDA, which is probably due its higher concentration compared to the other cannabinoids.

2.2.3. Reproducibility, R

The RSDs obtained in the reproducibility studies are shown in Table 1. The maximum RSD value was 2.13 for CBGA. The other cannabinoids show RSD values lower than 1.91, and the lowest of the RSDs was 0.09 for CBDA, which is probably due to the higher concentration of this cannabinoid.

2.2.4. Recovery

The tests were performed by using three different concentrations to test the recovery values in the linearity range of the method.

Quantities of CBD (4, 8, and 24 µg/mL) were added, thus assessing concentrations similar to, higher, and lower than those found in samples.

Recovery was determined according to this modality for CBD and was 84.92%.

An evaluation of recovery on all the compounds present in the sample was carried out by proceeding with a further extraction with 10 mL of methanol-chloroform on the sample residue after the usual extraction; in this extract, some cannabinoids were present, and indirectly the percentage of recovery was determined.

Table 1. Validation parameters of RP-HPLC-UV method.

Compound	R^2	[1] LOD (µg/mL)	[2] LOQ (µg/mL)	[3] LOD (µg/mL)	[4] LOQ (µg/mL)	Intraday (Repeatability) RSD	Interday (Repeatability) RSD	Reproducibility RSD	Recovery (%)
CBDA	0.9999	0.34	1.05	0.11	0.37	5.65	5.05	0.09	96.06
CBGA	0.9999	0.32	0.98	0.12	0.40	4.71	4.34	2.13	93.90
CBG	0.9995	0.62	1.87	0.13	0.45	3.34	2.83	0.91	94.60
CBD	0.9995	0.63	1.91	0.17	0.58	4.89	4.44	0.70	84.92
THCV	0.9989	0.95	2.87	0.15	0.49	-	-	N.d. *	N.d. *
CBN	0.9999	0.28	0.84	0.06	0.21	2.59	2.95	0.81	97.08
Δ9-THC	0.9981	1.25	3.79	0.15	0.50	3.05	3.22	0.13	99.69
Δ8-THC	0.9987	1.02	3.10	0.17	0.56	3.81	3.64	0.74	100
CBC	0.9999	0.29	0.88	0.11	0.36	5.3	4.78	0.89	98.68
THCA	0.9998	0.43	1.29	0.11	0.37	5.55	5.01	1.91	95.27

[1] Limit of detection (LOD) determined by the calibration curves (Instrumental LOD = $(3.3 \times \sigma)/m$). [2] Limit of quantification (LOQ) determined by the calibration curves (Instrumental LOQ = $(10 \times \sigma)/m$). [3] LOD determined by the signal-to-noise ratio (Instrumental LOD: S/N = 3). [4] LOQ determined by the signal-to-noise ratio (Instrumental LOQ: S/N = 10). * Not detectable.

The percentage of recovery values, as shown in Table 1, were higher than 84.92% and can be considered very satisfactory. In fact, considering CBD, the percentages are higher than those previously reported in the literature [5].

2.2.5. Detection Limit, LOD

The instrumental limit of detection was determined by the calibration curve, according to the formulas expressed in Section 3.6. The instrumental limit of detection (LOD) values obtained for CBDA and CBGA (Table 1) were lower, while those of CBG and CBD were comparable with similar methods described in literature [5,27]. Low LOD values were found also for the other cannabinoids (THCV, CBN, Δ-9 THC, Δ-8 THC, CBC, THCA), indicating that the method is sensitive.

2.2.6. Quantification Limit, LOQ

The instrumental limit of quantification was determined by a calibration curve, according to the formulas expressed in Section 3.6, considering that the signal-to-noise method is particularly useful to quantify the cannabinoids present at lower concentrations, such as THC. As reported for the LODs, the instrumental limit of quantification (LOQ) values obtained for CBDA and CBGA (Table 1) were also lower than those reported in the literature, while those for CBG and CBD were comparable with those of other methods described for similar procedures [5,27]. In addition, the other cannabinoids (THCV, CBN, Δ-9 THC, Δ-8 THC, CBC, THCA) showed low LOQs. The instrumental noise was registered in μV, by performing 3 blank injections with the ASTM method [28] given by the instrument, and a maximum CV% of 3.49% was calculated for all individual compounds to determine the single LOD and LOQ, which was considered acceptable.

2.2.7. Linearity

In order to evaluate the linearity of the method, eight different points of standard mixture solutions were analyzed in triplicate by HPLC-UV.

The following equations are related to the calibration curves in a concentration range between 0.01–100 μg/mL: CBDA, $y = 18955x - 1612.6$ ($r^2 = 0.9999$); CBGA, $y = 19796x - 3475.7$ ($r^2 = 0.9999$); CBG, $y = 18094x - 9195.3$ ($r^2 = 0.9995$); CBD, $y = 13703x - 6009.5$ ($r^2 = 0.9995$); THCV, $y = 18534x - 15213$ ($r^2 = 0.9989$); CBN, $y = 34148x - 7943.1$ ($r^2 = 0.9999$); Δ9 − THC, $y = 19893x - 31896$ ($r^2 = 0.9981$); Δ8-THC, $y = 17526x - 18267$ ($r^2 = 0.9987$); CBC, $y = 18590x - 4777.1$ ($r^2 = 0.9999$); THCA, $y = 18239x - 8969.3$ ($r^2 = 0.9998$) (Table 1).

With the aid of the equation obtained from the calibration curve, the quantity of each cannabinoid was calculated.

To express the data relative to the content of the individual cannabinoid as a percentage (%, p/p) referred to the dried material, it is necessary to refer to the weight of the sample considering the dilution factor. The linearity in the concentration range analyzed was good for cannabinoid standards, being $r^2 > 0.998$, as reported before.

2.3. *Cannabinoids in Hemp Samples*

The method developed in this study was applied to quali-quantitative analysis of main cannabinoids in two samples of hemp inflorescences. The samples analyzed, belonging to the same variety of *Cannabis sativa* L., did not show a significant difference in the concentration of the target compounds. As shown in Table 2, CBDA is the only cannabinoid for which a different concentration was determined. The other cannabinoids had a similar or the same concentration (e.g., CBGA, CBG, CBN, Δ-9-THC, and Δ-8-THC) in both samples. THCV was not found in the hemp inflorescence samples analyzed, as shown in Figure 2 and Table 2. Δ-9-THC and Δ-8-THC were found at a low concentration, below the legal limit. Under the current legislation regarding *Cannabis sativa* L. cultivation [6,29], in fact, the total content of THC must not be higher than 0.2% and in any case within 0.6%. Indeed, only the hemp varieties reported in the *Common catalogue of varieties of agricultural plant species* can be cultivated without authorization [6,7]. These kinds of results confirmed that the analyzed samples were correctly classified as hemp, since the quantity of Δ8-THC and Δ9-THC was found to be lower than the limits established by the legislation. According to what is indicated in literature [30], in the hemp variety considered (Futura 75), the most present compound was CBDA, followed by CBD; all the other compounds were in very low amounts ranging from 0.01 to 0.06%. CBGA is the compound from which all other cannabinoids are biosynthesized [5], which is probably why it was found at a low concentration in both samples examined.

The number of cannabinoids in hemp samples is reported in Table 2.

Table 2. Number of cannabinoids in hemp samples.

Sample	Cannabinoids									
	CBDA (%)	CBGA (%)	CBG (%)	CBD (%)	THCV (%)	CBN (%)	Δ9-THC (%)	Δ8-THC (%)	CBC (%)	THCA (%)
L11	0.88 ± 0.04	0.02 ± 0.00	0.02 ± 0.00	0.48 ± 0.02	N.d. *	0.01 ± 0.00	0.06 ± 0.00	0.03 ± 0.00	0.03 ± 0.00	0.03 ± 0.00
CV%	5.05	4.34	2.83	4.44		2.95	3.22	3.64	4.78	5.10
L5	0.93 ± 0.06	0.02 ± 0.00	0.02 ± 0.00	0.45 ± 0.03	N.d. *	0.01 ± 0.00	0.06 ± 0.00	0.03 ± 0.00	0.02 ± 0.00	0.04 ± 0.00
CV%	6.48	1.28	1.73	6.28		1.49	0.21	2.20	2.98	7.17

* Not detectable.

3. Materials and Methods

3.1. Chemicals, Standards and Apparatus

All chemicals used were of analytical grade. Methanol p.a CAS 67-56-1, chloroform p.a CAS 67-66-3, acetonitrile CAS 75-05-8, water CAS 7732-18-5, and orthophosphoric acid CAS 7664-38-2 were purchased from Sigma-Aldrich (St. Louis, MO, USA). Nitrogen, pure gas for analysis CAS 7727-37-9 was purchased from SIAD Spa (Bergamo, Italy). Standard mixture of phytocannabinoids 0.1% in acetonitrile: Cannabidiolic acid (0.01%) CAS 1244-58-2, cannabigerolic acid (0.01%) CAS 25555-57-1, cannabigerol (0.01%) CAS 25654-31-3, cannabidiol (0.01%) CAS 13956-29-1, tetrahydrocannabivarin (0.01%) CAS 31262-37-0, cannabinol (0.01%) CAS 521-35-7, tetrahydrocannabinolic acid (0.01%) CAS 23978-85-0, Δ-9-tetrahydrocannabinol (0.01%) CAS 1972-08-3, Δ-8-tetrahydrocannabinol (0.01%) CAS 5957-75-5, cannabichromene (0.01%) CAS Number 20675-51-8, were purchased from Cayman Chemical Company, (Ann Arbor, MI, USA). Cannabidiol 1.0 mg/mL in methanol CAS 13956-29-1: LGC Standards S.r.l., (Milan, Italy).

Analytical mill, IKA A11 Basic (IKA® Werke GMBH & Co. KG, Germany). Analytical balance with precision of 0.1 mg, mod. E42, (Gibertini, Italy). Vortex vibrating shaker, mod. ST5, (Janke & Kunkel, Germania). Centrifuge mod. ALC, PK 120 (Thermo Electron Corporation, Massachusetts, USA). Termoblock heating block, mod. A120, (Falc, Italy). Natural ventilation stove. Sieve with 1 mm meshes. Tilting shaker. Ultrasound bath Branson 2150, (Danbury-CT, USA). Volumetric flasks of 1, 2, 10 and 25 mL. SOVIREL-type tubes with screw cap. Glass syringes with luer lock attachment, 0.45 μm nylon membrane filters. Microsyringes from 1 to 1000 μL. HPLC Cannabis Analyzer for Potency Prominence-i LC-2030C equipped with a reverse phase C18 column, Nex-Leaf CBX Potency 150 × 4.6 mm, 2.7 μm with a guard column Nex-Leaf CBX 5 × 4.6 mm, 2.7, UV detector and acquisition software LabSolutions version 5.84 (Shimazu, Kyoto, Japan).

3.2. Sampling

The samples were supplied by a company that produces industrial hemp. In particular, two samples (L11 and L5) of inflorescences of *Cannabis sativa* L. Futura 75 were analyzed, having come from the same land and harvested in August 2017, and supplied by Enecta Srl. Sampling of material was carried out on a population of hemp plants, according to a systematic path, so that the sample taken was representative of the particle, excluding the edges, taking the upper third of the selected plant as indicated in Reg. (EU) No 1155/2017 [31]. The sample was dried in an oven at 35 °C ± 1 to constant weight, and gross wood parts and seeds with a length of more than 2 mm were removed. The samples were then subjected to grinding and subsequent sieving through a sieve with 1 mm meshes. The sieved material was transferred into polypropylene containers and stored under nitrogen atmosphere, protected from light at a temperature of −20 °C until extraction. Three independent replicates were performed for each sample, and three HPLC injections were performed for each replication.

3.3. *Cannabinoid Extraction*

To extract cannabinoids, an aliquot of powder sample, about 25 mg, was weighed using an analytical balance; 10 mL of methanol-chloroform extraction solvent 9:1 (*v*/*v*) was added as reported by De Backer et al. (2009) [32], Jin et al. (2017) [33], and was placed first for 10 min on an oscillating oscillator set at 350 oscillations per minute and then for 10 min in an ultrasonic bath. The sample was centrifuged for 10 min at 1125 g, and the supernatant was removed. The extraction was performed twice. The two fractions containing cannabinoids were collected in a 25 mL volumetric flask and were brought to volume with methanol/chloroform (9:1, *v*/*v*). The samples were filtered with a 45 µm nylon filter. Two mL of the filtered extract was transferred to a glass tube. The solvent was removed, leading to dryness with the help of a weak nitrogen flow, and recovered with 500 µL acetonitrile. The solution was injected into an HPLC-UV.

3.4. *Preparation of Standard Solution*

Appropriate aliquots of a standard mixture of cannabinoids are diluted with acetonitrile to obtain solutions of known concentration, in particular eight points in a concentration range between 0.05 and 100 µg/mL (0.05, 0.50, 4.17, 8.33, 16.70, 25.00, 50.00, 100.00 µg/mL). The standard solutions were prepared to construct calibration curves for the 10 cannabinoids considered: CBDA, CBGA, CBG, CBD, THCV, CBN, Δ9-THC, Δ8-THC, CBC, and THCA. The standard solutions were stored away from light at a temperature of −20 °C. The stability of standard solutions stored at −20 °C was evaluated every week for 3 months with the HPLC-UV system, and no degradation of cannabinoids was found.

3.5. *HPLC Conditions*

For the RP-HPLC analysis, the column was thermostated at 35 °C, and the autosampler was thermostated to 4 °C. Sample concentration was 4 mg/mL, and injection volume was 5.0 µL. UV detection was used at 220 nm, and gradient elution was used at flow rate of 1.6 mL/min according to the following procedure. Eluent mixture: Water + 0.085% phosphoric acid (A), acetonitrile + 0.085% phosphoric acid (B). Gradient elution: 70% of B up to 3 min, 85% of B to 7 min, 95% of B to 7.01 up to 8.00 min, and 70% of B up to 10 min. The eluent mixture was previously filtered with a Millipore system equipped with a 0.2 µm nylon filter.

3.6. *Validation Parameters*

3.6.1. Precision

Precision is the closeness of agreement among independent test results, obtained with stipulated conditions and usually in terms of standard deviation or relative standard deviation [34].

Precision was calculated with the following formula: $CV\% = [(SD/\bar{x}) \times 100]$, where SD is the estimate of the standard deviation and $\bar{x}$ is the average of the replications made.

3.6.2. Repeatability, R

The repeatability (intraday) of the method was evaluated by analyzing three replicates of the same sample, injected three times on the same day, performed by the same operator with the same method and instrument. The result corresponds to the arithmetic mean of the three determinations made considering the estimate of the standard deviation (SD) calculated on the three replicates performed.

The repeatability (interday) of the method was evaluated by performing three replicates of the same sample, injected three times on three different days, performed by the same operator with the same method and instrument. The result corresponds to the arithmetic mean of the three determinations made considering the estimate of the standard deviation (SD) calculated on the three replicates performed.

3.6.3. Reproducibility, R

Reproducibility was evaluated by the agreement between the results obtained on the same sample with the same procedure carried out by different operators in the laboratory and was measured with the coefficient of variation.

3.6.4. Recovery

Recovery is the fraction of analyte that was added to the sample being tested. Recovery was expressed as a percentage (R (%)) according to the following formula: R (%) = $[(Cf - C)/Cc] \times 100$, where Cf is the endogenous amount of the cannabinoid in the sample plus the amount of standard added to the analyte under examination. C is the endogenous amount present in the sample not added with the standard. Cc is the amount of the standard analyte added to the sample.

3.6.5. Detection Limit, LOD

The detection limit is the smallest amount or concentration of analyte in the sample that can be reliably distinguished from zero [34]. It can be calculated using the following formula: $LOD = (3.3 \times \sigma)/m$, where: σ represents the residual standard deviation of the calibration curve and m represents the slope of the calibration curve.

Furthermore, the LOD of the method from the signal (S)/noise (N) ratio can be determined as LOD: S/N = 3.

3.6.6. Quantification limit, LOQ

The quantification limit is the concentration of analyte below which it is determinable with a level of precision that is too low with inaccurate results. The LOQ can be determined according to the following formula: $LOQ = (10 \times \sigma)/m$, where σ represents the residual standard deviation of the calibration curve and m represents the slope of the calibration curve.

The LOQ of the method can also be determined by the signal-to-noise ratio (S/N): LOQ: S/N = 10.

3.6.7. Linearity

Linearity can be tested by examination of a plot of residuals produced by linear regression of the responses on the concentrations in an appropriate calibration set [34].

In order to quantify the analytes of interest, the equation of the calibration curve obtained for each standard is used. The equation is: $y = ax + b$, where y = area of the analyte obtained by HPLC/UV analysis, a = slope of the calibration curve, x = unknown concentration (μg/mL) of analyte in the sample, b = intercept of the calibration curve.

4. Conclusions

One of the most relevant problems in analytical determinations for quality control, especially when there are legal problems related with quantitation, such as for cannabis, relates to the proficiency of laboratories. Therefore, detailed and validated procedures that are freely available are essential for the full understanding of any analytical step and its careful application. This is also true for "daily" methods that can be easily applied for quality control, carried out using traditional RP-HPLC and UV-Vis detectors, with less efficient performance than diode-array detectors but with lower costs, rendering them affordable even for small laboratories.

The validated method described herein allows the quantitative determination of the 10 most relevant cannabinoids using a single wavelength (220 nm) in 8 min. A full separation is obtained, even in the elution sequence of a difficult resolution, of the group of peaks related to CBGA, CBG, CBD, and THCV (from 3.5 to 4.5 min).

The method is applied to cannabis inflorescences and involves extraction in methanol/ chloroform, drying of the extract, taking it up in acetonitrile and injection into an HPLC. The method has sensitivity and accuracy to discriminate samples with amounts of Δ-9- and Δ-8-THC (total THC content) that are below the limit of 0.2% from those that are subjected to legal restrictions in many EU countries, with a total THC content above 0.6%, which cannot be classified as hemp. Due to its simplicity and rapidity, it can be used to check raw material or crops during the harvesting period.

A detailed standard operating procedure (SOP), as a supplementary information file, is also available, so that any operator with basic knowledge of HPLC can easily apply it and make all the elution and calibration control checks using commercially available mixtures of standards, which are more affordable and sustainable than single cannabinoid standards in terms of costs and solvents used for calibration.

Supplementary Materials: The following are available online. File S1: Standard operating procedure (SOP) of the method presented in this article, Table S1: Calibration curves relating to the standard solution of 10 cannabinoids determined by RP-HPLC-UV method, Figure S1: Calibration curves relating to the standard solution of 10 cannabinoids determined by RP-HPLC-UV method, File S2: Preliminary tests carried out for development of the analytical procedure by RP-HPLC-UV.

Author Contributions: Conceptualization, T.G.T., M.M. and M.T.; Methodology, M.M.; Software, M.M. and S.S.; Validation, M.M.; Formal analysis, M.M.; Investigation, T.G.T. and M.M.; Resources, T.G.T. and M.M.; Data curation, M.M., T.G.T. and M.T.; Writing—original draft preparation, M.M., M.T. and T.G.T.; writing—review and editing, T.G.T. and S.S.; Visualization, T.G.T.; Supervision, T.G.T.; project administration, T.G.T.; funding acquisition, T.G.T.

Funding: This research received no external funding; this trial received financial support from Enecta Srl.

Acknowledgments: The authors gratefully acknowledge Enecta Srl for providing samples. The experimentation was conducted in the context of a PhD project entitled *"Harmonized procedures of analysis of medical, herbal, food and industrial cannabis: development and validation of cannabinoids' quality control methods, of extraction and preparation of derivatives from the plant raw material, according to the product destination"* and funded by ENECTA Srl.

Conflicts of Interest: The authors declare no conflict of interest.

References

1. Montserrat-de la Paz, S.; Marín-Aguilar, F.; García-Gimenez, M.D.; Fernández-Arche, M.A. Hemp (*Cannabis sativa* L.) seed oil: Analytical and phytochemical characterization of the unsaponifiable fraction. *J. Agric. Food Chem.* **2014**, *62*, 1105–1110. [CrossRef] [PubMed]
2. Appendino, G.; Chianese, G.; Taglialatela-Scafati, O. Cannabinoids: Occurrence and medicinal chemistry. *Curr. Med. Chem.* **2011**, *18*, 1085–1099. [CrossRef]
3. Andre, C.M.; Hausman, J.F.; Guerriero, G. *Cannabis sativa*: The plant of the thousand and one molecules. *Front. Plant. Sci.* **2016**, *7*, 19. [CrossRef] [PubMed]
4. Aizpurua-Olaizola, O.; Soydaner, U.; Öztürk, E.; Schibano, D.; Simsir, Y.; Navarro, P.; Usobiaga, A. Evolution of the cannabinoid and terpene content during the growth of *Cannabis sativa* plants from different chemotypes. *J. Nat. Prod.* **2016**, *79*, 324–331. [CrossRef]
5. Brighenti, V.; Pellati, F.; Steinbach, M.; Maran, D.; Benvenuti, S. Development of a new extraction technique and HPLC method for the analysis of non-psychoactive cannabinoids in fibre-type *Cannabis sativa* L.(hemp). *J. Pharm. Biomed. Anal.* **2017**, *143*, 228–236. [CrossRef] [PubMed]
6. Legge 2 Dicembre 2016, n.242. Disposizioni per la Promozione della Coltivazione e della Filiera Agroindustriale della Canapa (16G00258), GU Serie Generale n. 304 del 30-12-2016. Available online: https://www.gazzettaufficiale.it/eli/id/2016/12/30/16G00258/sg (accessed on 1 June 2019).
7. European Commission (EC). European Union Common Catalogue of Varieties of Agricultural Plant Species, Plant Variety Database. Available online: http://ec.europa.eu/food/plant/plant_propagation_material/plant_variety_catalogues_databases/search//public/index.cfm (accessed on 1 June 2019).

8. Pisanti, S.; Malfitano, A.M.; Ciaglia, E.; Lamberti, A.; Ranieri, R.; Cuomo, G.; Laezza, C. Cannabidiol: State of the art and new challenges for therapeutic applications. *Pharmacol. Ther.* **2017**, *175*, 133–150. [CrossRef] [PubMed]
9. Mechoulam, R.; Gaoni, Y. Recent advances in the chemistry of hashish. In *The Chemistry of Organic Natural Products/Progrès dans la Chimie des Substances Organiques Naturelles*; Springer International Publishing: Vienna, Austria, 1967; pp. 175–213.
10. Radwan, M.M.; Wanas, A.S.; Chandra, S.; ElSohly, M.A. Natural Cannabinoids of Cannabis and Methods of Analysis. In Cannabis sativa *L.-Botany and Biotechnology*, 1st ed.; Springer International Publishing AG: Cham, Switzerland, 2017; pp. 161–182.
11. Leghissa, A.; Hildenbrand, Z.L.; Schug, K.A. A review of methods for the chemical characterization of cannabis natural products. *J. Sep. Sci* **2018**, *41*, 398–415. [CrossRef] [PubMed]
12. Citti, C.; Pacchetti, B.; Vandelli, M.A.; Forni, F.; Cannazza, G. Analysis of cannabinoids in commercial hemp seed oil and decarboxylation kinetics studies of cannabidiolic acid (CBDA). *J. Pharm Biomed. Anal.* **2018**, *149*, 532–540. [CrossRef] [PubMed]
13. Citti, C.; Braghiroli, D.; Vandelli, M.A.; Cannazza, G. Pharmaceutical and biomedical analysis of cannabinoids: A critical review. *J. Pharm Biomed. Anal.* **2018**, *147*, 565–579. [CrossRef] [PubMed]
14. Rodrigues, A.; Yegles, M.; Van Elsué, N.; Schneider, S. Determination of cannabinoids in hair of CBD rich extracts consumers using gas chromatography with tandem mass spectrometry (GC/MS–MS). *Forensic. Sci. Int.* **2018**, *292*, 163–166. [CrossRef]
15. Cardenia, V.; Gallina Toschi, T.; Scappini, S.; Rubino, R.C.; Rodriguez Estrada, M.T. Development and validation of a Fast gas chromatography/mass spectrometry method for the determination of cannabinoids in *Cannabis sativa* L. *J. Food Drug Anal.* **2018**, *26*, 1283–1292. [CrossRef] [PubMed]
16. Leghissa, A.; Hildenbrand, Z.L.; Foss, F.W.; Schug, K.A. Determination of cannabinoids from a surrogate hops matrix using multiple reaction monitoring gas chromatography with triple quadrupole mass spectrometry. *J. Sep. Sci* **2018**, *41*, 459–468. [CrossRef] [PubMed]
17. Patel, B.; Wene, D.; Fan, Z.T. Qualitative and quantitative measurement of cannabinoids in cannabis using modified HPLC/DAD method. *J. Pharm Biomed. Anal.* **2017**, *146*, 15–23. [CrossRef] [PubMed]
18. Burnier, C.; Esseiva, P.; Roussel, C. Quantification of THC in Cannabis plants by fast-HPLC-DAD: A promising method for routine analyses. *Talanta* **2019**, *192*, 135–141. [CrossRef] [PubMed]
19. Pellati, F.; Brighenti, V.; Sperlea, J.; Marchetti, L.; Bertelli, D.; Benvenuti, S. New Methods for the Comprehensive Analysis of Bioactive Compounds in *Cannabis sativa* L.(hemp). *Molecules* **2018**, *23*, 2639. [CrossRef]
20. Ciolino, L.A.; Ranieri, T.L.; Taylor, A.M. Commercial cannabis consumer products part 2: HPLC-DAD quantitative analysis of cannabis cannabinoids. *Forensic. Sci. Int.* **2018**, *289*, 438–447. [CrossRef]
21. Fekete, S.; Sadat-Noorbakhsh, V.; Schelling, C.; Molnár, I.; Guillarme, D.; Rudaz, S.; Veuthey, J.L. Implementation of a generic liquid chromatographic method development workflow: Application to the analysis of phytocannabinoids and *Cannabis sativa* extracts. *J. Pharm. Biomed. Anal.* **2018**, *155*, 116–124. [CrossRef]
22. Purschke, K.; Heinl, S.; Lerch, O.; Erdmann, F.; Veit, F. Development and validation of an automated liquid-liquid extraction GC/MS method for the determination of THC, 11-OH-THC, and free THC-carboxylic acid (THC-COOH) from blood serum. *Anal. Bioanal. Chem.* **2016**, *408*, 4379–4388. [CrossRef]
23. Pacifici, R.; Marchei, E.; Salvatore, F.; Guandalini, L.; Busardò, F.P.; Pichini, S. Evaluation of cannabinoids concentration and stability in standardized preparations of cannabis tea and cannabis oil by ultra-high performance liquid chromatography tandem mass spectrometry. *Clin. Chem. Lab. Med.* **2017**, *55*, 1555–1563. [CrossRef]
24. Casiraghi, A.; Roda, G.; Casagni, E.; Cristina, C.; Musazzi, U.M.; Franzè, S.; Gambaro, V. Extraction Method and Analysis of Cannabinoids in Cannabis Olive Oil Preparations. *Planta Med.* **2018**, *84*, 242–249. [CrossRef]
25. Lin, S.Y.; Lee, H.H.; Lee, J.F.; Chen, B.H. Urine specimen validity test for drug abuse testing in workplace and court settings. *J. Food Drug Anal.* **2018**, *26*, 380–384. [CrossRef] [PubMed]
26. Mudge, E.M.; Murch, S.J.; Brown, P.N. Leaner and greener analysis of cannabinoids. *Anal. Bioanal. Chem.* **2017**, *409*, 3153–3163. [CrossRef] [PubMed]

27. Gul, W.; Gul, S.W.; Radwan, M.M.; Wanas, A.S.; Mehmedic, Z.; Khan, I.I.; ElSohly, M.A. Determination of 11 cannabinoids in biomass and extracts of different varieties of Cannabis using high-performance liquid chromatography. *J. AOAC Int.* **2015**, *98*, 1523–1528. [CrossRef] [PubMed]
28. ASTM. *ASTM E685-93(2013), Standard Practice for Testing Fixed-Wavelength Photometric Detectors Used in Liquid Chromatography*; ASTM International: West Conshohocken, PA, USA, 2013. Available online: www.astm.org (accessed on 1 June 2019).
29. Regulation (EU) No 1307/2013 of the European Parliament and of the Council of 17 December 2013 Establishing Rules for Direct Payments to Farmers under the Support Schemes within the Framework of the Common Agricultural Policy and Repealing Council Regulation (EC) No 637/2008 and Council Regulation (EC) No 73/2009. Available online: https://eur-lex.europa.eu/legal-content/EN/TXT/?uri=celex:32013R1307 (accessed on 1 June 2019).
30. del M. Contreras, M.; Jurado-Campos, N.; Sánchez-Carnerero Callado, C.; Arroyo-Manzanares, N.; Fernández, L.; Casano, S.; Ferreiro-Vera, C. Thermal desorption-ion mobility spectrometry: A rapid sensor for the detection of cannabinoids and discrimination of *Cannabis sativa* L. chemotypes. *Sens. Actuators B Chem.* **2018**, *273*, 1413–1424. [CrossRef]
31. Commission Delegated Regulation (EU) 2017/1155 of 15 February 2017 Amending Delegated Regulation (EU) No 639/2014 as Regards the Control Measures Relating to the Cultivation of Hemp, Certain Provisions on the Greening Payment, the Payment for Young Farmers in Control of a Legal Person, the Calculation of the per Unit Amount in the Framework of Voluntary Coupled Support, the Fractions of Payment Entitlements and Certain Notification Requirements Relating to the Single Area Payment Scheme and the Voluntary Coupled Support, and Amending Annex X to Regulation (EU) No 1307/2013 of the European Parliament and of the Council. Available online: https://eur-lex.europa.eu/eli/reg_del/2017/1155/oj (accessed on 1 June 2019).
32. De Backer, D.B.; Debrus, B.; Lebrun, P. Innovative development and validation of an HPLC/DAD method for the qualitative and quantitative determination of major cannabinoids in cannabis plant material. *J. Chromatogr. B Anal. Technol. Biomed. Life Sci.* **2009**, *877*, 4115–4124. [CrossRef] [PubMed]
33. Jin, D.; Jin, S.; Yu, Y.; Lee, C.; Chen, J. Analytical & Bioanalytical Techniques Classification of Cannabis Cultivars Marketed in Canada for Medical Purposes by Quantification of Cannabinoids and Terpenes Using HPLC-DAD and GC-MS. *J. Anal. Bioanal. Tech.* **2017**, *8*, 1–9.
34. Thompson, M.; Ellison, S.L.; Wood, R. Harmonized guidelines for single-laboratory validation of methods of analysis (IUPAC Technical Report). *Pure Appl. Chem.* **2002**, *74*, 835–855. [CrossRef]

Sample Availability: Samples of fiber-type hemp extracts are available from the authors.

Article

Determination of Flavonoid Glycosides by UPLC-MS to Authenticate Commercial Lemonade

Ying Xue [1,2,3,†], Lin-Sen Qing [2,†], Li Yong [3], Xian-Shun Xu [3], Bin Hu [3], Ming-Qing Tang [1,2] and Jing Xie [1,*]

1 School of Pharmacy, Chengdu Medical College, Chengdu 610500, China
2 Chengdu Institute of Biology, Chinese Academy of Sciences, Chengdu 610041, China
3 Sichuan Provincial Center for Disease Control and Prevention, Chengdu 610041, China
* Correspondence: xiejing@cmc.edu.cn; Tel.: +86-28-6230-8658
† These authors contributed equally to this work.

Academic Editors: Marcello Locatelli, Angela Tartaglia, Dora Melucci, Abuzar Kabir, Halil Ibrahim Ulusoy and Victoria Samanidou
Received: 31 July 2019; Accepted: 16 August 2019; Published: 20 August 2019

Abstract: So far, there is no report on the quality evaluation of lemonade available in the market. In this study, a sample preparation method was developed for the determination of flavonoid glycosides by ultra-performance liquid chromatography–mass spectrometry (UPLC-MS) based on vortex-assisted dispersive liquid-liquid microextraction. First, potential flavonoids in lemonade were scanned and identified by ultra-performance liquid chromatography–time of flight mass spectrometry (UPLC-TOF/MS). Five flavonoid glycosides were identified as eriocitrin, narirutin, hesperidin, rutin, and diosmin according to the molecular formula provided by TOF/MS and subsequent confirmation of the authentic standard. Then, an ultra-performance liquid chromatography–triple quadrupole mass spectrometry (UPLC-QqQ/MS) method was developed to determine these five flavonoid glycosides in lemonade. The results showed that the content of rutin in some lemonade was unreasonably high. We suspected that many illegal manufacturers achieved the goal of low-cost counterfeiting lemonade by adding rutin. This suggested that it was necessary for relevant departments of the state to make stricter regulations on the quality standards of lemonade beverages.

Keywords: vortex-assisted dispersive liquid-liquid microextraction; flavonoid glycoside; UPLC-MS; counterfeiting lemonade

---

1. Introduction

Lemon (*Citrus limon* L.) is considered the third most important citrus species in the world [1], with a large spectrum of biological activities that include antioxidant, antimicrobial, antiviral, antifungal, and antidiabetic activities [2,3], generating a large variety of healthy foods. Flavonoids are widely contained in lemon, conferring the typical taste and biological activities to lemon. According to the aglycone structures, flavonoids are divided into four classes: flavanones, flavones, flavonols, and flavans. Flavanones are the most abundant flavonoids, which are usually present in the 7-*O*-diglycoside form. Lemon flavanones are present in glycoside or aglycone forms. Among the phytochemicals, hesperetin and eriodictyol are the most abundant types of aglycones and rutinoside is the most abundant types of glycoside forms [4,5]. It has been reported that hesperidin and eriocitrin were the most abundant flavonoids in all the lemon juices studied and far exceed others [6–8].

Due to the high cost of fruit, counterfeiting of fruit juice has become a common problem in the industry. The three most common forms of counterfeiting are: (1) When a kind of cheaper fruit is used to replace all or part of it, (2) when a monomeric compound contained in the fruit with another cheaper source is added, and (3) when it is completely made up of additives such as artificial sweeteners, preservatives, and colors [9]. As the products produced with the first two counterfeiting methods contain some natural characteristic ingredients, they can generally meet the national testing standards [10]. However, such kinds of counterfeit juice not only seriously affect consumer confidence in the juice market, but may also cause a series of food safety problems. In addition to pure lemon juice, lemonade containing lemon ingredients occupies an increasing market share in the beverage market. Thus, it is of great scientific significance and commercial value to identify the authenticity of lemonade available in the market.

Some methods for analyzing lemon juice have been reported, such as nuclear magnetic resonance [11], $^{13}C/^{12}C$ isotope ratios [12], capillary electrochromatography (CEC) [13], and HPLC [6,7,14]. Among them, HPLC was considered as the most reliable method for determining flavonoids with high selectivity and sensitivity. Lemonade beverages currently available in the market contain a large number of additives besides a small amount of lemon juice. Therefore, a new sample preparation method is required to selectively separate and enrich low-content flavonoids from lemonade, so as to identify the authenticity of lemonade.

At present, sample preparation methods of flavonoids can be divided into liquid-liquid extraction (LLE) and solid phase extraction (SPE) [15–17]. However, they have some inherent disadvantages. For example, LLE needs a substantial amount of toxic solvents and is time-consuming. SPE materials are expensive and have poor reusability [18]. The dispersive liquid-liquid microextraction (DLLME) method developed in recent years can make up for these disadvantages [19–21]. DLLME can not only separate and enrich target analyte from aqueous solution, but also reduce or even eliminate the matrix interference of samples. Therefore, DLLME is considered to be an effective pretreatment method for food samples with the advantages of less solvent consumption, simple operation, high enrichment factor, etc. In order to improve the work efficiency by speeding up the mass transfer process and reducing the balance time, some assistant emulsification methods were also applied to improve the performance of DLLME, such as ultrasound-assisted [22], vortex-assisted [23], air-assisted [24], and microwave-assisted [25] DLLME. Currently, there are some studies on sample preparation of flavonoids by DLLME. However, as far as we know, there is no research on flavonoids in lemonade.

In this work, the sample preparation of flavonoids in lemonade was firstly performed by the vortex-assisted dispersive liquid-liquid microextraction (VA-DLLME) method. Then, the structure and content of flavonoids in lemonade available on the market from eight different manufacturers were identified and determined by ultra-performance liquid chromatography–time of flight mass spectrometry (UPLC-TOF/MS) and ultra-performance liquid chromatography–triple quadrupole mass spectrometry (UPLC-QqQ/MS), respectively. Finally, the counterfeiting phenomenon of lemonade was evaluated according to the determination results of flavonoids. As far as we know, this study was the first determination of flavonoid glycosides by UPLC-MS to authenticate commercial lemonade available in the market.

2. Results and Discussion

2.1. Identification of Flavonoid Glycosides by UPLC-TOF/MS

The time of flight mass spectrometer (TOF MS) was used to scan and identify potential flavonoids in lemonade for the first time in this work. As one of the most common high-resolution MS, TOF MS can determine the exact molecular formula of the target compound, thus identifying the structure in a complex matrix. After the target compound was located and identified, the triple quadrupole mass spectrometer (QqQ MS) was an excellent choice for subsequent quantitative analysis [26].

In this study, according to the calculation based on the molecular formula by TOF and the subsequent confirmation of the authentic standard under the same chromatographic conditions, 5 flavonoid glycosides in lemonade available in the market were located and identified (Figure 1), which were eriocitrin, narirutin, hesperidin, rutin, and diosmin, respectively. As shown in Table 1, the error of each compound in high-resolution MS is within ±5 ppm, which is the acceptable error limit for structure confirmation [27].

rutinose-O

	R_1	R_2
Eriocitrin	OH	H
Narirutin	H	H
Hesperidin	OH	OCH_3

	R_3	R_4	R_5
Rutin	OH	O-rutinose	OH
Diosmin	O-rutinose	H	OCH_3

Figure 1. Chemical structures of eriocitrin, narirutin, hesperidin, rutin, and diosmin.

Table 1. UPLC-MS parameters of five analytes in the negative ion-scan mode.

Analyte	TOF/MS			QqQ/MS	
	Quasi-Molecular Ion (m/z)	Error (ppm)	Product Ion (m/z)	Parent Ion (m/z)	Product Ion (m/z)
eriocitrin	595.16788	1.7	287.0586, 151.0065	595	287 *, 151 #
narirutin	579.17238	0.8	271.0612	579	271 *, 151 #
hesperidin	609.18386	2.2	301.0737	609	301 *, 286 #
rutin	609.14689	1.3	301.0383, 300.0281	609	300 *, 271 #
diosmin	607.16784	1.0	299.0582, 284.0345	607	299 *, 284 #

Note: * quantitative ion, # qualitative ion.

2.2. The Selection of VA-DLLME Conditions

Since the extraction conditions have a crucial influence on the performance of VA-DLLME, single-factor experiments were carried out to select the extraction conditions of the amount of ethyl acetate and acetonitrile. In the present study, recoveries of 5 flavonoid glycosides were assessed by means of fixing one variable and changing the other two variables. The results are shown in Figure 2. Due to structural differences, the recoveries of the 5 flavonoid glycosides were different, but the overall trend was relatively consistent. Based on the investigation of single-factor experiments, the VA-DLLME condition was set as 1 mL of lemonade, 500 μL acetonitrile, and 1.5 mL ethyl acetate.

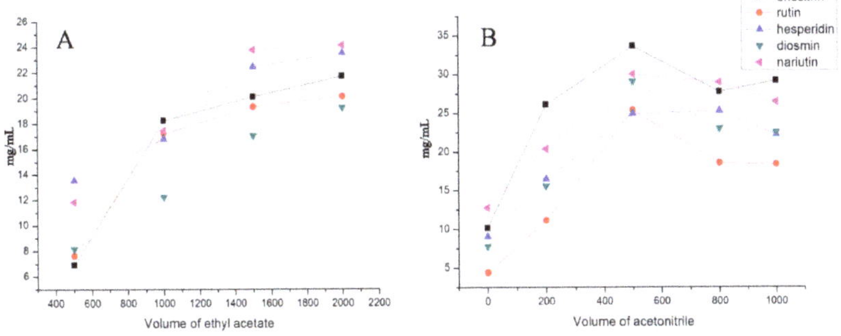

Figure 2. The evaluation of extraction conditions of the amount of ethyl acetate (**A**) and acetonitrile (**B**).

2.3. Determination of Flavonoid Glycosides by UPLC–QqQ/MS

All 5 flavonoid glycosides are acidic compounds. Therefore, acid mobile phase could increase the separating degree, symmetry factor, and the number of theoretical plates. Considering the ion suppression induced by a high concentration of acid, 0.2% formic acid was finally added into the mobile phase [28]. In order to optimize the MS condition of 5 flavonoid glycosides in the present study, all of these target analytes were tested in direct infusion mode using the full-scan MS method, respectively. It was found that the negative mode was more sensitive and selective than the positive mode. By optimizing mass spectrum variables, including the vaporizer temperature, sheath gas pressure, aux gas pressure, the parent/product ion pairs, collision energy, and S-Lens value, two stable product ions with high sensitivity were selected for MRM analysis (Table 1). The representative mass spectra of lemonade samples are shown in Figure 3.

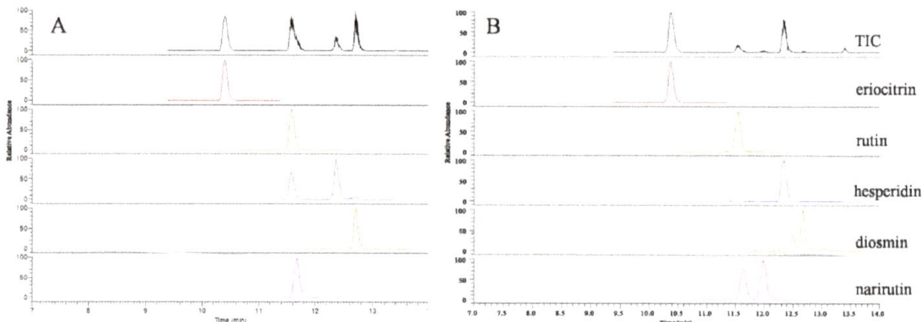

Figure 3. UPLC–QqQ/MS total ion count chromatograms of five flavonoid glycosides standards (**A**) and commercial lemonade sample (**B**).

2.4. Speculation on the Possible Counterfeiting Means of Lemonade

A total of 8 batches of lemonade samples purchased from local supermarkets was determined by the proposed UPLC-QqQ/MS method. The contents of 5 flavonoid glycosides are shown in Table 2. The content of total flavonoid glycosides in lemonade varies greatly. On the surface, it seems that the higher the content of total flavones, the higher the amount of lemon juice added in lemonade, which means the better the quality of the product. However, after further analysis of the content of monomeric compounds, it was found that the main ingredients in S1–S4 were flavanone glycosides (mainly hesperidin and eriocitrin) and the content of flavonol glycosides (mainly rutin) was relatively low. This result is consistent with the distribution characteristics of flavonoid glycosides in *Citrus* L. With regard to S5–S8, the content of rutin is extremely high and hesperidin as a characteristic ingredient of *Citrus* L. is not detected (nd). Hesperidin was the predominant flavonoid glycoside in lemon reported by the previous study. For example, Mannan et al. reported values of 67 ± 15 mg/L for hesperidin in 38 natural lemon juices, showing that the absence of this compound in lemonade shows it to be a possible counterfeit [29]. Under normal circumstances, the content of rutin in lemon should not exceed the content of hesperidin. Xi reported the contents of hesperidin and rutin in juice varied from 105.5 to 210.3 μg/g and nd to 3.82 μg/g, respectively [30]. Due to the abnormal phenomenon in our work, we have reason to suspect that S5–S8 were counterfeited as there was no or only a small trace of lemon juice and had instead a large amount of rutin added to meet the national testing standards (colorimetric assay by UV-Vis) of fruit juice products. Rutin is widely distributed in the plant kingdom. It was reported that its content in *Sophora japonica* L. was up to 37.8% [31]. Therefore, only with a simple separation process the commercialized low-cost supply of rutin can be realized [32]. For example, the price of rutin reagent supplied by Aladdin is ¥368/100 g and if it is a crude extract of food-grade, the price will be even lower. According to the testing method of total flavonoids in fruit juice beverage specified by national standard, rutin also has an obvious response in the colorimetric assay by UV-Vis at a wavelength of 420 nm. Therefore, illegal businessmen achieved the goal of low-cost counterfeiting lemonade by adding rutin.

Table 2. The contents of five flavonoid glycosides in eight lemonade samples (μg/100 mL).

Sample No.	Eriocitrin	Rutin	Hesperidin	Diosmin	Narirutin	Total
S1	0.04	0.27	1.00	nd	nd	1.31
S2	0.92	0.34	3.82	0.66	nd	5.75
S3	1.73	0.29	16.33	2.95	nd	21.30
S4	28.96	6.01	28.30	0.28	0.74	49.07
S5	2.66	191.54	nd	0.64	nd	194.83
S6	0.04	243.71	nd	nd	nd	243.75
S7	0.05	264.24	nd	nd	nd	264.29
S8	0.35	470.00	nd	nd	nd	470.35

3. Material and Methods

3.1. Chemicals and Reagents

A total of eight lemonade samples were purchased from local supermarkets. A total of five authentic standards of eriocitrin, narirutin, hesperidin, rutin, and diosmin were obtained from Chengdu Push Bio-technology Co., Ltd. (Chengdu, China). The Milli-Q water purification system was used to prepare ultra-pure water for UPLC analysis (Millipore, Bedford, MA, USA). Formic acid and acetonitrile of LC/MS grade for UPLC-MS analysis were purchased from Sigma-Aldrich. Ethyl acetate, ether, dichloromethane, methanol, acetone, and acetonitrile of analytical grade were purchased from Sinopharm Chemical Reagent Co., Ltd. (Shanghai, China).

3.2. Preparation of Standard Solution

Stock solutions of five target analytes (eriocitrin, narirutin, hesperidin, rutin, and diosmin) were prepared by dissolving each 10 mg authentic standard in 10 mL of methanol. Then, 250 μL of each of the five stock solutions was transferred to a 50-mL volumetric flask and diluted with 20% methanol to obtain the mixed stock solution. Next, 500 μL of mixed stock solution was transferred to a 50-mL volumetric flask and diluted with 20% methanol to obtain the working solution I with a concentration of approximately 50 ng/mL. Finally, mixed working solutions II–V were obtained by diluting working solution I with respective concentrations of about 20.0 ng/mL, 10.0 ng/mL, 5.0 ng/mL, and 2 ng/mL. All the solutions were stored in a refrigerator at 4 °C before use.

3.3. Sample Preparation by the VA-DLLME Procedure

Accurately add 1 mL of lemonade to a 4 mL centrifuge tube, then add 500 μL of acetonitrile and 1.5 mL ethyl acetate, then vortex for 30 s. After centrifugation, the upper organic phase was transferred. The extraction was repeated once using another 1.5 mL of ethyl acetate and the combined solvent of the upper organic phase was removed by a Termovap Sample Concentrator. The resulting residue was re-dissolved in 1 mL of 20% methanol and filtered through a 0.22 μm filter for UPLC-MS analysis.

3.4. UPLC–MS Analysis

3.4.1. Identification of Flavonoid Glycosides by UPLC-TOF/MS

The Shimadzu UPLC ((Shimadzu, Kyoto, Japan) system consists of an online degasser (DGU-20A5R), an auto-sampler (SIL-30AC), two pumps (LC-30AD), and a column oven (CTO-30aHE). Chromatographic separation was performed on a Waters BEHC18 analytical column (2.1 × 100 mm, 1.7 μm, Waters, Milford, MA, USA) at 40 °C. The mobile phase consisted of 0.2% formic acid and acetonitrile. The linear gradient elution with a constant flow rate of 0.2 mL/min was 10%~10%~40%~95%~10% acetonitrile at

0~1~10~13~15 min. The sample solution and mixed working solutions of 5 μL were injected into the UPLC system by the auto-sampler.

TOF/MS measurements in negative ion mode were performed on a 4600 Q-TOF mass spectrometer (AB Sciex, Concord, CA, USA) equipped with an electrospray ionization (ESI) source with the following parameters: Ion source gas 1 (GS1) at 50 psi, ion source gas 1 (GS1) (N_2) at 50 psi, curtain gas at 35 psi, temperature at 500 °C, and ionspray voltage floating at −4500 V. The mass range was set to *m*/*z* 100–800. The system was operated under Analyst 1.6 and Peak 2.0 (AB Sciex, Concord, CA, USA) and used an APCI negative calibration solution to calibrate the instrument's mass accuracy in real-time.

3.4.2. Determination of Flavonoid Glycosides by UPLC-QqQ/MS

Chromatographic separation was the same as that used in UPLC-TOF/MS analysis described above. QqQ/MS measurements in negative ion mode were accomplished by a triple quadrupole mass spectrometer equipped with an ESI source (Thermo Fisher Scientific, San Jose, CA, USA). The determination of the target analytes was performed in a multi-reaction monitoring mode. The MS parameters were as follows: Vaporizer temperature and capillary temperature both 350 °C, aux gas pressure of 10 Arb, sheath gas pressure of 40 Arb, ion sweep gas pressure of 2 Arb, discharge current of 4.0 μA, and spray voltage of −2000 V. Data collection and processing were conducted with Thermo Xcalibur Workstation (Version 2.2, Thermo).

3.5. Analytical Figures of Merit

Method validation was performed according to the above UPLC–QqQ/MS conditions. After it was determined by the mixed working solutions I–V, the calibration curves of five analytes were obtained as shown in Table 3 by taking the concentration of each authentic standard as the abscissa (x) and the corresponding peak area as the ordinate (y), respectively. The limit of detection (LOD) and the limit of quantification (LOQ) were measured by a gradual dilution process of the standard stock solutions until the signal-to-noise ratio of 3:1 and 10:1, respectively. The precision was evaluated by standard working solution III, which was tested within one day to determine the intra-day precision and was tested within 3 days to determine the inter-day precision. The repeatability was evaluated by analyzing six independent portions of sample S4 with parallel running. The recovery was carried out by spiking an amount of about 1:1 of authentic standards to six independent portions of sample S4 with parallel running. The validation results are summarized in Table 3, which show that the present developed UPLC–QqQ/MS method meets the requirements of quantitative analysis and was appropriate for the determination of five flavonoid glycosides in lemonade. The analytical figures of merit were compared with those of several other quantitative methods reported for flavonoid glycosides in lemon as shown in Table 4.

Table 3. The results of method validation.

Analyte	Regression Equation	Linear Range (ng/mL)	LOD	LOQ	Precision (RSD, n = 6)		Repeatability (n = 6)		Recovery (n = 6)	
	(y = ax + b, r^2)		(ng/mL)	(ng/mL)	Intra-Day	Inter-Day	Mean (µg/100 mL)	RSD	Mean	RSD
eriocitrin	y = 523.81x − 82.61, 0.996	2.01–50.3	0.70	2.01	1.36%	3.53%	28.9	3.22%	88.5%	3.93%
rutin	y = 1034.77x − 1160.22, 0.996	2.44–60.9	0.81	2.44	2.51%	3.79%	6.01	4.62%	89.9%	4.51%
hesperidin	y = 513.03x + 252.28, 0.998	2.01–50.4	0.70	2.10	1.97%	2.58%	28.3	3.47%	88.7%	5.30%
diosmin	y = 769.76x + 84.74, 0.995	2.14–53.4	0.71	2.14	2.02%	3.17%	0.28	5.47%	102%	2.61%
narirutin	y = 556.25x + 148.69, 0.997	2.02–51.2	0.70	2.02	1.48%	3.69%	0.74	4.86%	92.8%	4.47%

Table 4. Comparison of analytical methods reported for determination of flavonoid glycosides in lemon.

Method	Analyte	Linea Range	LOD	LOQ	Recovery
CEC [13]	eriocitrin, narirutin, hesperidin	5–200 µg/mL	2.5 µg/mL	5 µg/mL	71–112%
HPLC/UV [33]	narirutin, hesperidin, diosmin	0.25–20 µg/mL	-	0.1 µg/mL	-
HPLC/UV [34]	narirutin	2–50 mg/L	1.25 mg/L	2.5 mg/L	83%
	hesperidin	2–50 mg/L	1.0 mg/L	2.5 mg/L	74%
HPLC/UV [35]	eriocitrin	1.01–50.50 µg/mL	0.02 µg/mL	0.065 µg/mL	103.10%
	narirutin	0.505–10.10 µg/mL	0.024 µg/mL	0.18 µg/mL	99.14%
	hesperidin	5.00–100.00 µg/mL	0.04 µg/mL	0.132 µg/mL	99%
	rutin	0.101–10.100 µg/mL	0.079 µg/mL	0.263 µg/mL	98.37%
UPLC/UV [36]	eriocitrin	0.5–130 mg/L	6 µg/kg	-	90.50%
	narirutin	0.05–300 mg/L	5 µg/kg	-	87.40%
	hesperidin	0.05–500 mg/L	8 µg/kg	-	92.70%
	rutin	0.05–310 mg/L	5 µg/kg	-	88.40%
	diosmin	0.01–200 mg/L	8 µg/kg	-	100.80%

4. Conclusions

In this study, five flavonoid glycosides of eriocitrin, narirutin, hesperidin, rutin, and diosmin in lemonade were identified and determined by UPLC-TOF/MS and UPLC-QqQ/MS, respectively. By estimating the content characteristics of flavonoid glycosides in the samples, we highly suspected that some lemonade available in the market was counterfeited: Cheap rutin was added to increase the content of "total flavonoids of lemon". This indicates that besides using total flavonoids, the content of multiple flavonoid compounds should be included in the quality standard of lemonade in the future.

Author Contributions: Conceptualization, L.-S.Q. and J.X.; Data curation, Y.X.; Formal analysis, Y.X. and M.-Q.T.; Funding acquisition, L.-S.Q.; Methodology, Y.X.; Project administration, J.X.; Resources, L.Y., B.H., and X.-S.X.; Supervision, L.Y., B.H., X.-S.X., and J.X.; Validation, Y.X.; Writing—original draft, L.-S.Q.; Writing—review & editing, J.X.

Funding: Financial support was provided by the West Light Foundation of The Chinese Academy of Sciences (No. 2018XBZG_XBQNXZ_A_005).

Conflicts of Interest: The authors declare no conflict of interest.

References

1. Papoutsis, K.; Pristijono, P.; Golding, J.B.; Stathopoulos, C.E.; Scarlett, C.J.; Bowyer, M.C.; Vuong, Q.V. Impact of different solvents on the recovery of bioactive compounds and antioxidant properties from lemon (*Citrus limon* L.) pomace waste. *Food Sci. Biotechnol.* **2016**, *25*, 971–977. [CrossRef] [PubMed]
2. Kim, J.; Jayaprakasha, G.K.; Patil, B.S. Lemon (*Citrus lemon* L. Burm) as a source of unique bioactive compounds. *Acta Hortic.* **2014**, *1040*, 377–380. [CrossRef]
3. Proteggente, A.R.; Pannala, A.S.; Paganga, G.; Buren, L.v.; Wagner, E.; Wiseman, S.; Put, F.v.d.; Dacombe, C.; Rice-Evans, C.A. The antioxidant activity of regularly consumed fruit and vegetables reflects their phenolic and vitamin C composition. *Free Radic. Res.* **2002**, *36*, 217–233. [CrossRef] [PubMed]
4. Tripoli, E.; Guardia, M.L.; Giammanco, S.; Majo, D.D.; Giammanco, M. Citrus flavonoids: Molecular structure, biological activity and nutritional properties: A review. *Food Chem.* **2007**, *104*, 466–479. [CrossRef]
5. Del Río, J.A.; Fuster, M.D.; Gómez, P.; Porras, I.; García-Lidón, A.; Ortuño, A. Citrus limon: A source of flavonoids of pharmaceutical interest. *Food Chem.* **2004**, *84*, 457–461. [CrossRef]
6. Abad-García, B.; Garmón-Lobato, S.; Sánchez-Ilárduya, M.B.; Berrueta, L.A.; Gallo, B.; Vicente, F.; Alonso-Salces, R.M. Polyphenolic contents in Citrus fruit juices: Authenticity assessment. *Eur. Food Res. Technol.* **2014**, *238*, 803–818. [CrossRef]
7. Abad-García, B.; Berrueta, L.A.; Garmón-Lobato, S.; Urkaregi, A.; Gallo, B.; Vicente, F. Chemometric characterization of fruit juices from spanish cultivars according to their phenolic compound contents: I. Citrus fruits. *J. Agric. Food Chem.* **2012**, *60*, 3635–3644. [CrossRef]
8. Kawaii, S.; Tomono, Y.; Katase, E.; Ogawa, K.; Yano, M. Quantitation of flavonoid constituents in citrus fruits. *J. Agric. Food Chem.* **1999**, *47*, 3565–3571. [CrossRef]
9. Zhang, M.; Li, X.X.; Jia, H.F.; Wang, X. The research on detection techniques of adulteration about fruit juice in China. *Food Res. Dev.* **2016**, *37*, 205–208.
10. National Standards of China. *General Analytical Methods for Beverage*; National Standards of China: GB/T 12143-2008; Standards Press of China: Beijin, China, 2009.
11. Salazar, M.O.; Pisano, P.L.; González Sierra, M.; Furlan, R.L.E. NMR and multivariate data analysis to assess traceability of argentine citrus. *Microchem. J.* **2018**, *141*, 264–270. [CrossRef]
12. Guyon, F.; Auberger, P.; Gaillard, L.; Loublanches, C.; Viateau, M.; Sabathié, N.; Salagoïty, M.H.; Médina, B. $^{13}C/^{12}C$ isotope ratios of organic acids, glucose and fructose determined by HPLC-co-IRMS for lemon juices authenticity. *Food Chem.* **2014**, *146*, 36–40. [CrossRef] [PubMed]
13. Desiderio, C.; De Rossi, A.; Sinibaldi, M. Analysis of flavanone-7-O-glycosides in citrus juices by short-end capillary electrochromatography. *J. Chromatogr. A* **2005**, *1081*, 99–104. [CrossRef] [PubMed]
14. Vaclavik, L.; Schreiber, A.; Lacina, O.; Cajka, T.; Hajslova, J. Liquid chromatography–mass spectrometry-based metabolomics for authenticity assessment of fruit juices. *Metabolomics* **2012**, *8*, 793–803. [CrossRef]

15. Qing, L.S.; Xiong, J.; Xue, Y.; Liu, Y.M.; Guang, B.; Ding, L.S.; Liao, X. Using baicalin-functionalized magnetic nanoparticles for selectively extracting flavonoids from *Rosa chinensis*. *J. Sep. Sci.* **2011**, *34*, 3240–3245. [CrossRef] [PubMed]
16. Qing, L.S.; Xue, Y.; Liu, Y.M.; Liang, J.; Xie, J.; Liao, X. Rapid magnetic solid-phase extraction for the selective determination of isoflavones in soymilk using baicalin-functionalized magnetic nanoparticles. *J. Agric. Food Chem.* **2013**, *61*, 8072–8078. [CrossRef] [PubMed]
17. Qing, L.S.; Xue, Y.; Zhang, J.G.; Zhang, Z.F.; Liang, J.; Jiang, Y.; Liu, Y.M.; Liao, X. Identification of flavonoid glycosides in Rosa chinensis flowers by liquid chromatography–tandem mass spectrometry in combination with ^{13}C nuclear magnetic resonance. *J. Chromatogr. A* **2012**, *1249*, 130–137. [CrossRef] [PubMed]
18. Andrade-Eiroa, A.; Canle, M.; Leroy-Cancellieri, V.; Cerdà, V. Solid-phase extraction of organic compounds: A critical review. part II. *TrAC Trends Anal. Chem.* **2016**, *80*, 655–667. [CrossRef]
19. Mousavi, L.; Tamiji, Z.; Khoshayand, M.R. Applications and opportunities of experimental design for the dispersive liquid–liquid microextraction method—A review. *Talanta* **2018**, *190*, 335–356. [CrossRef] [PubMed]
20. Sajid, M.; Alhooshani, K. Dispersive liquid-liquid microextraction based binary extraction techniques prior to chromatographic analysis: A review. *TrAC Trends Anal. Chem.* **2018**, *108*, 167–182. [CrossRef]
21. Rykowska, I.; Ziemblińska, J.; Nowak, I. Modern approaches in dispersive liquid-liquid microextraction (DLLME) based on ionic liquids: A review. *J. Mol. Liq.* **2018**, *259*, 319–339. [CrossRef]
22. Homem, V.; Alves, A.; Alves, A.; Santos, L. Ultrasound-assisted dispersive liquid–liquid microextraction for the determination of synthetic musk fragrances in aqueous matrices by gas chromatography–mass spectrometry. *Talanta* **2016**, *148*, 84–93. [CrossRef] [PubMed]
23. Xue, Y.; Xu, X.S.; Yong, L.; Hu, B.; Li, X.D.; Zhong, S.H.; Li, Y.; Xie, J.; Qing, L.S. Optimization of vortex-assisted dispersive liquid-liquid microextraction for the simultaneous quantitation of eleven non-anthocyanin polyphenols in commercial blueberry using the multi-objective response surface methodology and desirability function approach. *Molecules* **2018**, *23*, 2921.
24. Li, G.; Row, K.H. Air assisted dispersive liquid–liquid microextraction (AA-DLLME) using hydrophilic–hydrophobic deep eutectic solvents for the isolation of monosaccharides and amino acids from kelp. *Anal. Lett.* **2019**, 1–15. [CrossRef]
25. Mahmoudpour, M.; Mohtadinia, J.; Mousavi, M.M.; Ansarin, M.; Nemati, M. Application of the microwave-assisted extraction and dispersive liquid–liquid microextraction for the analysis of PAHs in smoked rice. *Food Anal. Methods* **2017**, *10*, 277–286. [CrossRef]
26. Chen, C.; Xue, Y.; Li, Q.M.; Wu, Y.; Liang, J.; Qing, L.S. Neutral loss scan - based strategy for integrated identification of amorfrutin derivatives, new peroxisome proliferator-activated receptor gamma agonists, from *Amorpha Fruticosa* by UPLC-QqQ-MS/MS and UPLC-Q-TOF-MS. *J. Am. Soc. Mass Sp.* **2018**, *29*, 685–693. [CrossRef] [PubMed]
27. Gross, M.L. Accurate masses for structure confirmation. *J. Am. Soc. Mass Sp.* **1994**, *5*, 57. [CrossRef]
28. Xie, J.; Li, J.; Liang, J.; Luo, P.; Qing, L.S.; Ding, L.S. Determination of contents of catechins in oolong teas by quantitative analysis of multi-components via a single marker (QAMS) method. *Food Anal. Methods* **2017**, *10*, 363–368. [CrossRef]
29. Hajimahmoodi, M.; Moghaddam, G.; Mousavi, S.; Sadeghi, N.; Oveisi, M.; Jannat, B. Total antioxidant activity, and hesperidin, diosmin, eriocitrin and quercetin contents of various lemon juices. *Trop. J. Pharm. Res.* **2014**, *13*, 951–956. [CrossRef]
30. Xi, W.; Lu, J.; Qun, J.; Jiao, B. Characterization of phenolic profile and antioxidant capacity of different fruit part from lemon (*Citrus limon* Burm.) cultivars. *J. Food Sci. Technol.* **2017**, *54*, 1108–1118. [CrossRef] [PubMed]
31. Tan, J.; Li, L.Y.; Wang, J.R.; Ding, G.; Xu, J. Study on quality evaluation of *Flos Sophorae* Immaturus. *Nat. Prod. Res. Dev.* **2018**, *30*, 138–146+174.
32. Chua, L.S. A review on plant-based rutin extraction methods and its pharmacological activities. *J. Ethnopharmacol.* **2013**, *150*, 805–817. [CrossRef] [PubMed]
33. Kanaze, F.I.; Gabrieli, C.; Kokkalou, E.; Georgarakis, M.; Niopas, I. Simultaneous reversed-phase high-performance liquid chromatographic method for the determination of diosmin, hesperidin and naringin in different citrus fruit juices and pharmaceutical formulations. *J. Pharm. Biomed. Anal.* **2003**, *33*, 243–249. [CrossRef]
34. Belajová, E.; Suhaj, M. Determination of phenolic constituents in citrus juices: Method of high performance liquid chromatography. *Food Chem.* **2004**, *86*, 339–343. [CrossRef]

35. Tu, X.; Yang, S.; Wu, Z.; Wu, C.; Zhang, L.; Lü, X. Simultaneous determination of eleven flavonoids in different lemon (*Citrus limon*) varieties by HPLC. *J. Hunan Agric. Univ.* **2016**, *42*, 543–548.
36. Zheng, J.; Zhao, Q.Y.; Zhang, Y.H.; Jiao, B.N. Simultaneous determination of main flavonoids and phenolic acids in citrus fruit by ultra performance liquid chromatography. *Sci. Agric. Sin.* **2014**, *47*, 4706–4717.

Sample Availability: Samples of the compounds are available from the authors.

Communication

Effects of Harvest Time on Phytochemical Constituents and Biological Activities of *Panax ginseng* Berry Extracts

Seung-Yeap Song [1,†], Dae-Hun Park [2,†], Seong-Wook Seo [3], Kyung-Mok Park [4], Chun-Sik Bae [5], Hong-Seok Son [6], Hyung-Gyun Kim [7], Jung-Hee Lee [7], Goo Yoon [1], Jung-Hyun Shim [1], Eunok Im [3], Sang Hoon Rhee [8], In-Soo Yoon [3,*] and Seung-Sik Cho [1,*]

1 Department of Pharmacy, College of Pharmacy, Mokpo National University, Jeonnam 58554, Korea; tgb1007@naver.com (S.-Y.S.); gyoon@mokpo.ac.kr (G.Y.); s10004jh@gmail.com (J.-H.S.)
2 Department of Nursing, Dongshin University, Jeonnam 58245, Korea; dhj1221@hanmail.net
3 Department of Pharmacy, College of Pharmacy, Pusan National University, Busan 46241, Korea; sswook@pusan.ac.kr (S.-W.S.); eoim@pusan.ac.kr (E.I.)
4 Department of Pharmaceutical Engineering, Dongshin University, Jeonnam 58245, Korea; parkkm@dsu.ac.kr
5 College of Veterinary Medicine, Chonnam National University, Gwangju 61186, Korea; csbae210@chonnam.ac.kr
6 School of Korean Medicine, Dongshin University, Jeonnam 58245, Korea; hsson@dsu.ac.kr
7 Department of Research Planning, Mokpo Marine Food-industry Research Center, Jeonnam 58621, Korea; khg8279@naver.com (H.-G.K.); bluebabyi@nate.com (J.-H.L.)
8 Department of Biological Sciences, Oakland University, Rochester, MI 48309, USA; srhee@oakland.edu
* Correspondence: insoo.yoon@pusan.ac.kr (I.-S.Y.); sscho@mokpo.ac.kr (S.-S.C.); Tel.: +82-51-510-2806 (I.-S.Y.), +82-61-450-2687 (S.-S.C.); Fax: +82-51-513-6754 (I.-S.Y.), +82-61-450-2689 (S.-S.C.)
† These authors contributed equally to this work.

Academic Editors: Marcello Locatelli, Angela Tartaglia, Dora Melucci, Abuzar Kabir, Halil Ibrahim Ulusoy and Victoria Samanidou
Received: 28 August 2019; Accepted: 12 September 2019; Published: 13 September 2019

Abstract: Ginseng (*Panax ginseng*) has long been used as a traditional medicine for the prevention and treatment of various diseases. Generally, the harvest time and age of ginseng have been regarded as important factors determining the efficacy of ginseng. However, most studies have mainly focused on the root of ginseng, while studies on other parts of ginseng such as its berry have been relatively limited. Thus, the aim of this study iss to determine effects of harvest time on yields, phenolics/ginsenosides contents, and the antioxidant/anti-elastase activities of ethanol extracts of three- and four-year-old ginseng berry. In both three- and fourfour-year-old ginseng berry extracts, antioxidant and anti-elastase activities tended to increase as berries ripen from the first week to the last week of July. Liquid chromatography-tandem mass spectrometry analysis has revealed that contents of ginsenosides except Rg1 tend to be the highest in fourfour-year-old ginseng berries harvested in early July. These results indicate that biological activities and ginsenoside profiles of ginseng berry extracts depend on their age and harvest time in July, suggesting the importance of harvest time in the development of functional foods and medicinal products containing ginseng berry extracts. To the best of our knowledge, this is the first report on the influence of harvest time on the biological activity and ginsenoside contents of ginseng berry extracts.

Keywords: ginseng berry; harvest time; ginsenoside; antioxidant activity; anti-elastase activity

1. Introduction

Ginseng (*Panax ginseng*) has long been used as a traditional medicine for the prevention and treatment of various diseases, including cancer, diabetes, inflammation, allergy, and cardiovascular diseases, in the East Asia, particularly in Korea and China [1,2]. Ginseng is one of the best known and most recognized medicinal herbs. Its pharmacological effects have been successfully demonstrated by numerous studies worldwide [3]. However, most studies have focused mainly on the root of ginseng, while studies on other parts of ginseng such as its berry and leaf are relatively limited [4].

More than sixty different ginsenosides have been identified from various parts of ginseng [2]. In particular, ginseng berry is known to have a distinct phytochemical profile. It contains significantly higher ginsenoside content than ginseng root [5,6]. Oral bioavailability of ginsenosides is generally very low. It is only 0.64% for Rb1 and 3.29% for Rg1 in rats [7,8]. However, oral absorption of ginsenoside Re is significantly higher (by 1.18–3.95 fold) after oral ingestion of a ginseng berry extract than pure ginsenoside Re [9]. To date, a few in vitro and in vivo studies have reported a variety of biological activities of ginseng berry on cancer, diabetes, sexual dysfunction, skin whitening, immunity, and liver injury. These studies are summarized in Table 1 [3,4,10–16]. A randomized and placebo-controlled clinical trial has also been performed to evaluate the safety and efficacy of ginseng berry extract on glycemic control [17].

Table 1. Chemical constituents and pharmacological activities of ginseng berry extracts reported in previous literature.

Ext. Solvent	Constituent	Activity	Region	Effective Dose (mg/kg) (route/animal/day)	Estimated Human Dose (mg/60 kg/day)	Ref.
Ethanol Water	Rb1, Rb2, Rd,Re,Rf, Rg1, Rg2, 20SRg3, Rg6, Rh1, Rh4,Rk1,Rk3, F1,F4	Hepatoprotective	South Korea	100–500 (PO/rat)	972.4–4862	[3]
Ethanol	Polysaccharide K	Anti-immunosenescent		30 (PO/mouse)	146	[4]
Butanol	Re	Antidiabetic	China	150 as ext.5–20 as Re(PO/mouse)	729 as ext.24.3–97.3 as Re	[10]
ND	Polysaccharides	Antidiabetic	USA	150 (IP/mouse)		[11]
Ethylacetate	Re	Antidiabetic	South Korea	20–50 (PO/mouse)	97.3–243.3	[12]
70% ethanol	Rb1, Rb2, Rc, Rd, Re, Rg1, Rg2	Penile erection	South Korea	20–150 (PO/rat)	194.5–1458.7	[13]
70% ethanol		Antipigmentation		In vitro		[14]
Butanol	Rg1, Re, Rh1, Rg2, Rb1, Rc, Rb2, Rb3, Rd, Rg3, 20R-Rg3, Rh2	Anticancer	USA	50 (PO/mouse)	243.3	[15]
Water	Rb1, Rb2, Rc, Rd, Re, Rf	Blood circulation		50–150 (PO/rat)	486.2–1458.7	[16]

A recent study reported the alterations of metabolomes during five different ginseng berry maturation stages and their effects on the functional bioactive compounds in ginseng [18]. Thus, information regarding the optimal harvest time of ginseng berry is needed to standardize the collection and pretreatment process of the plant material for its further development as functional foods or medicinal preparations. However, only a few studies have reported the influence of harvest time on the chemical and biological properties of ginseng berry. In a previous study, five different flower and berry development stages (flower bud, flowering, early berry, green berry, and red berry) were tested with respect to ginsenoside biosynthetic gene expression and ginsenoside contents in biochemical and molecular aspects [19]. However, we focused on the effect of harvest time on biological activities and chemical profiles of green-to-red ginseng berries, which is more relevant to the agricultural and industrial aspects. Here, the objective of the present study is to determine the effects of harvest time on yields, phenolic contents, ginsenoside contents, antioxidant activity, and the elastase inhibitory activity of ethanol extracts of three and four years old ginseng berry.

2. Results and Discussion

2.1. Drying and Extraction Yields of Ginseng Berry Extracts

To date, there have been no reports on yields related to the preparation of ginseng berry extracts. As shown in Table 2, after drying the harvested ginseng berry with hot air, the yield of this process ranged from 29.9% to 34.8% (31.8% on average). After extracting the dried ginseng berry with 70% ethanol, the yield of this process ranged from 8.8% to 12.6% (mean: 10.8%). The overall production yield of the extract was calculated to be 3.4%.

Table 2. Drying and extraction yields of ginseng berry extracts.

Sample	Drying (%, *w/w*)	Extraction (%, *w/w*)
3Y1W	29.7	11.2
3Y2W	31.2	11.0
3Y3W	34.8	8.8
3Y4W	32.2	11.9
3Y5W	31.3	11.4
4Y1W	32.0	10.4
4Y2W	33.1	10.8
4Y3W	32.8	11.2
4Y4W	30.7	9.2
4Y5W	29.9	12.6

2.2. Antioxidant Properties of Ginseng Berry Extracts

Antioxidant properties of ginseng berry extracts were assessed by measuring DPPH radical scavenging activity, reducing power, and total phenolic contents. DPPH antioxidant assay is a fast and easy method to evaluate free radical scavenging capacity of a given sample [20]. As shown in Figure 1, DPPH radical scavenging activities of three-year-old ginseng berry extracts tended to increase from 26.8% to 62.5% when the harvest time was delayed. Those of four-year-old ginseng berry extracts also showed similar tendency of increase from 11.0% to 72.7%. Extracts of four-year old ginseng berry harvested in the 3rd and 4th weeks of July exhibited DPPH radical scavenging activity comparable to the positive control (vitamin C), which tended to be higher than other groups (Figure 1). As shown in Figure 2, the reducing power tended to increase as the harvest time was delayed from the 3rd year 1st week (3Y1W) to 4th year 5th week (4Y5W). Extracts of four-year-old ginseng berry harvested in the 3rd week of July exhibited significantly higher reducing power than other groups (Figure 2). As shown in Figure 3, total phenol contents of three-year-old ginseng berry extracts tended to increase from 13.6% to 29.7% as the harvest time was delayed from 1st week to 5th week of July. Those of four-year-old ginseng berry extracts showed a similar tendency, increasing from 3.2% to 13.6%. Extracts of three-year-old ginseng berry harvested in the 4th week of July exhibited significantly higher reducing power than other groups (Figure 2). Although the temporal changes of mean DPPH activity tended to be roughly similar to those of mean total phenols, the harvest time to exhibit the highest DPPH activity (4Y3W and 4Y4W) was different from that for total phenols (3Y4W). This discrepancy could be attributed to other antioxidant phytochemicals besides phenols in the ginseng berry extract, which warrants further investigation.

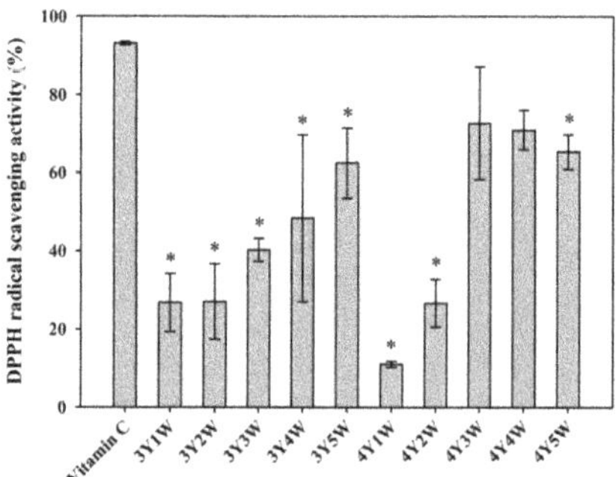

Figure 1. DPPH radical scavenging activity of ginseng berry extracts harvested at various time points. Rectangular bars and their error bars represent means and standard deviations, respectively ($n = 3$). The '*m*Y*n*W' on the x-axis means *m*-year-old ginseng berry harvested in the *n*th week of July. *, significantly lower than the 'Vitamin C' group (positive control).

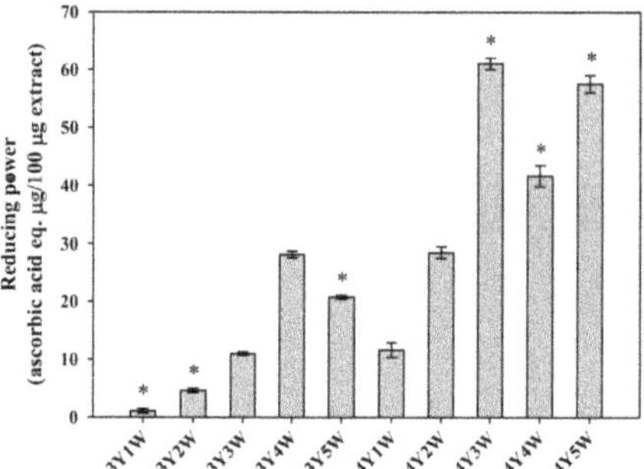

Figure 2. Reducing power of ginseng berry extracts harvested at various time points. Rectangular bars and their error bars represent means and standard deviations, respectively ($n = 3$). The '*m*Y*n*W' on the x-axis means *m*-year-old ginseng berry harvested in the *n*th week of July. *, significantly different from other groups.

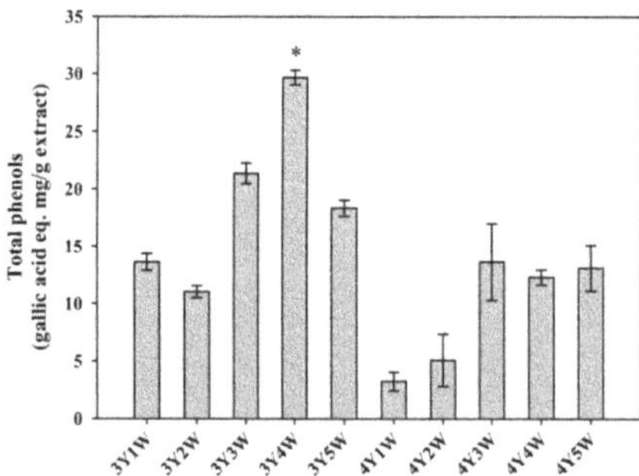

Figure 3. Total phenolic contents of ginseng berry extracts harvested at various time points. Rectangular bars and their error bars represent means and standard deviations, respectively ($n = 3$). The '*m*Y*n*W' on the x-axis means *m*-year-old ginseng berry harvested in the *n*th week of July. *, significantly different from other groups.

2.3. Elastase Inhibitory Activity of Ginseng Berry Extracts

Figure 4 shows inhibitory effects of ginseng berry extracts on elastase activity. Elastase inhibitory activities of three-year-old ginseng berry extracts tended to increase from 32.5% to 70.0% as the harvest time was delayed from 1st week to 5th week of July. Those of four-year-old ginseng berry extracts showed a similar tendency, increasing from 43.2% to 84.6%. Extracts of three-year-old and four-year-old ginseng berry harvested in the 3rd, 4th, and 5th weeks of July exhibited significantly higher inhibitory activities than the phosphoramidon group (as positive control).

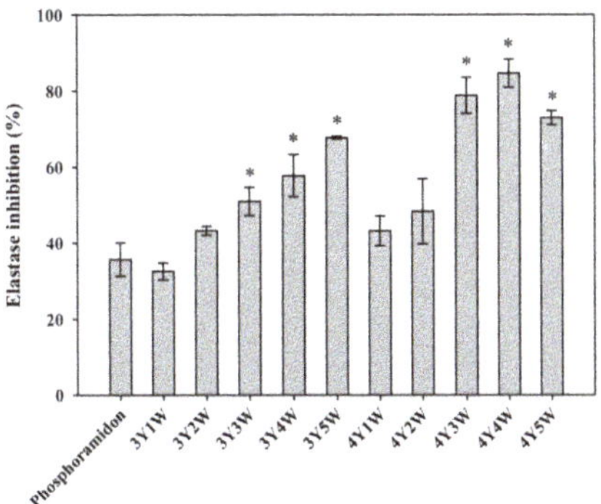

Figure 4. Elastase inhibitory activities of ginseng berry extracts harvested at various time points. Rectangular bars and their error bars represent means and standard deviations, respectively ($n = 3$). The 'mYnW' on the x-axis means m-year-old ginseng berry harvested in the nth week of July. *, significantly higher than the 'phosphoramidon' group as positive control.

2.4. Contents of Ginsenosides in Ginseng Berry Extracts

Contents of ginsenosides Rb3, Rc, Rd, Re, and Rg1 in ginseng berry extracts were determined by LC-MS/MS analysis. Typical mass chromatograms are shown in Figure 5. Contents of five ginsenosides in extracts of ginseng berry harvested at various times are shown in Figure 6. As shown in Figure 6B,C, Rc and Rd contents were significantly higher in extracts of four-year-old ginseng berry harvested in the 1st week of July than those in other groups. Similarly, Rb3 and Re contents tended to be the highest in extracts of four-year-old ginseng berry harvested in early July (Figure 6A,D). However, Rg1 content exhibited a slightly different tendency from other ginsenosides. It tended to be the highest in extracts of three-year-old ginseng berry harvested in the 4th week of July and four-year-old ginseng berry harvested in the 2nd week of July (Figure 6E). Contents of all ginsenosides studied were the lowest in extracts of four-year-old ginseng berry harvested in the last week of July.

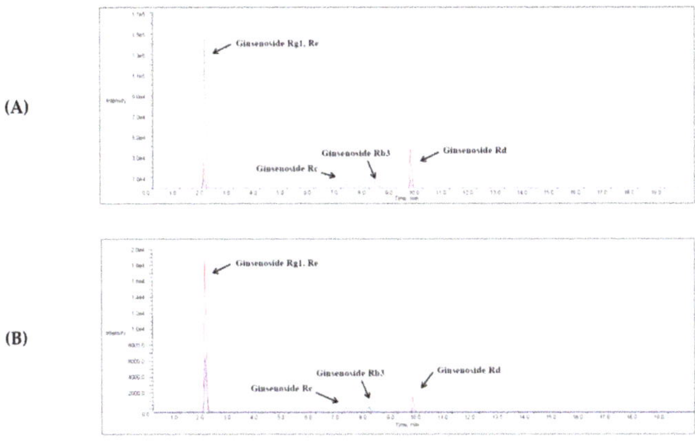

Figure 5. *Cont.*

(C)

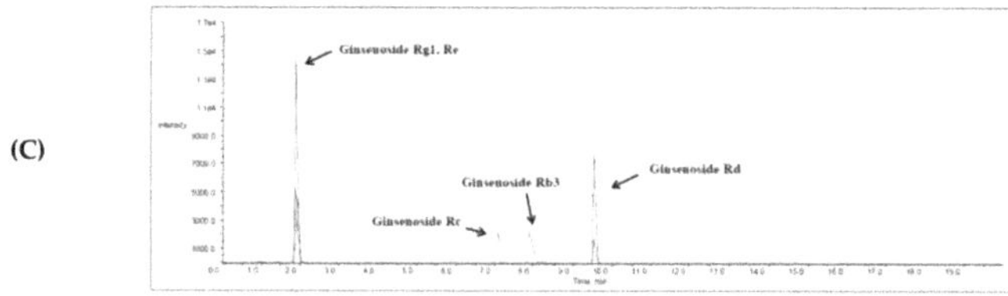

Figure 5. Representative chromatograms of ginsenosides Rb3, Rc, Rd, Re, and Rg1 in calibration standard (**A**), three-year-old ginseng berry extract sample (**B**), and four-year-old ginseng berry extract sample (**C**).

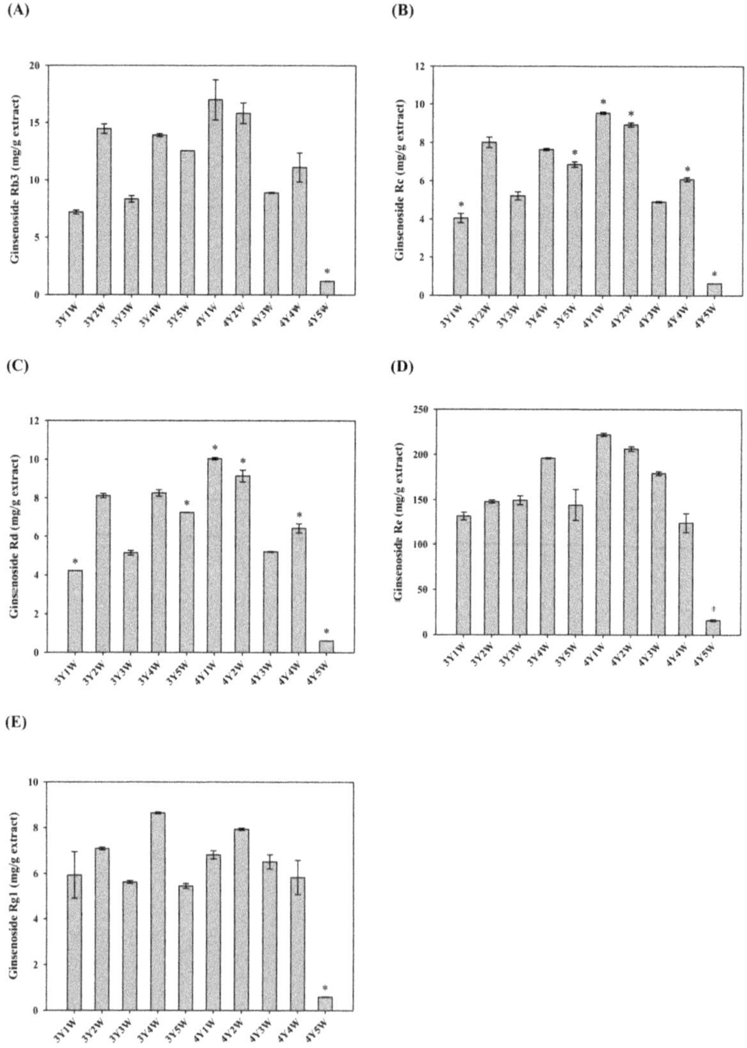

Figure 6. Contents of ginsenosides Rb3 (**A**), Rc (**B**), Rd (**C**), Re (**D**), and Rg1 (**E**) in ginseng berry extracts harvested at various time points. Rectangular bars and their error bars represent means and standard deviations, respectively (n = 3). The 'mYnW' on the x-axis means m-year-old ginseng berry harvested in the nth week of July. *, significantly different from other groups.

2.5. Effects of Harvest Time on Chemical Constituents and Biological Activities of Ginseng Berry Extracts

In both 3- and four-year-old ginseng berry extracts, antioxidant (DPPH radical scavenging activity and reducing power) and anti-elastase activities tended to increase as berries ripened from the first week to the last week of July. However, contents of ginsenosides except Rg1 tended to be higher in four-year-old ginseng berries harvested in early July than those in other groups. These results indicate that biological activities and ginsenoside profiles of ginseng berry extracts depend on their age and harvest time in July, suggesting a need to optimize harvest time for the development of functional foods and medicinal products containing ginseng berry extracts. To the best of our knowledge, this is the first study to report the impact of harvest time on antioxidant and anti-elastase activities as well as ginsenoside contents of ginseng berry extracts.

3. Materials and Methods

3.1. Plant Materials

Ginseng berry was harvested from three-year-old and four-year-old ginseng cultivated in a local farm (Healthy Sam-Farm, Jeonbuk, Korea) every week from July 1 to July 30, 2017. Dried ginseng berry of 25 g was extracted with 70% ethanol at room temperature for 72 h. After removing ethanol, residual water part was freeze-dried and then stored at −70 °C before analysis.

3.2. DPPH Free Radical Assay

Antioxidant activity was determined with 2,2-diphenyl-1-picrylhydrazyl (DPPH) radical scavenging assay. Briefly, 1 mL sample solution (final concentration: 1–20 mg/mL; dissolved in DDW) was added to 0.4 mM DPPH sample solution (1 mL; dissolved in methanol) and then vortex-mixed. The resultant mixture was allowed to react at room temperature in the dark for 10 min. Its absorbance at 517 nm was then measured using a microplate reader (Perkin Elmer, Waltham, MA, USA). DPPH free radical scavenging activities of samples in terms of their IC_{50} (μg/mL) values were evaluated. Vitamin C was used as a positive control.

3.3. Reducing Power

Reducing power was determined using a modified reducing power assay. Briefly, sample (0.1 mL) was added to 0.2 M sodium phosphate buffer (0.5 mL) and 1% potassium ferricyanide (0.5 mL), followed by incubation at 50 °C for 20 min. Subsequently, 10% trichloroacetic acid solution (0.5 mL) was added to the reaction mixture followed by centrifugation at 12,000× *g* for 10 min. The supernatant was mixed with distilled water (0.5 mL) and 0.1% iron (III) chloride solution (0.1 mL). The absorbance of the resulting solution was measured at 700 nm. Reducing powers of samples are expressed as vitamin C equivalents [21].

3.4. Determination of Total Phenolic Content

Total phenolic content was determined by Folin–Ciocalteu assay. Briefly, 1 mL sample (final concentration: 5 mg/mL) was mixed with 1 mL of 2% sodium carbonate solution and 1 mL of 10% Folin–Ciocalteu's phenol reagent. After incubating the mixture at room temperature for 10 min, its absorbance was measured at 750 nm using microplate reader and compared with the calibration curve of gallic acid. Data are expressed as milligrams of gallic acid equivalents per gram of sample [21].

3.5. Determination of Elastase Inhibitory Activity

Elastase inhibitory activity was determined as previously described [22]. Briefly, 10 μL elastase derived from porcine pancreas (10 μg/mL) was mixed with 90 μL of 0.2 M Tris-HCl, 100 μL of STANA (2.5 mM, *N*-Succinyl-Ala-Ala-Ala-*p*-nitroanilide), and 50 μL of the sample and incubated at 37 °C for 30 min. The reaction mixture was then centrifuged at 15,000× *g* for 10 min to obtain supernatant. The absorbance of the supernatant was measured at 405 nm using a microplate reader. Phosphoramidon, an inhibitor of elastase from *Pseudomonas aeruginosa*, was used as a positive control.

3.6. Determination of Ginsenoside Contents

Contents of ginsenosides Rb3, Rc, Rd, Re, and Rg1 were determined by high-performance liquid chromatography-tandem mass spectrometry (LC-MS/MS) analysis. The LC-MS/MS system consisted of a Sciex HPLC system coupled with a triple quadrupole mass spectrometer (Triple Quad 4500, AB Sciex, Framingham, MA, USA). The mobile phase for the HPLC system consisted of water containing 0.1% formic acid (solvent A) and acetonitrile containing 0.1% formic acid (solvent B). It was eluted at 0.4 mL/min. A gradient elution protocol was used: solvent A:solvent B, *v/v* ramped from 72:28 to 65:35 for 6 min; ramped from 65:35 to 0:100 for 4 min; held at 0:100 for 1 min; back to 72:28 for 4 min; and then held at 72:28 form 5 min. Chromatographic separation was performed using a reversed-phase column ZORBAX Eclipse Plus (C18, 3 × 100 mm, particle size 1.8 μm; Agilent, Santa Clara, CA, USA), which was maintained at 40 °C. To avoid contamination by particles, the mobile phase was filtered through a 0.45 μm filter device (PEEK, Supelco, Taufkirchen, Germany) before use. The mass spectrometer was operated in the positive ion mode using multiple reaction monitoring (MRM). The following ion source parameters were used: temperature, 600 °C; collision gas pressure, 9 mTorr; sheath gas pressure, 40 Arb; and auxiliary valve flow rate, 10 Arb. Detailed mass spectrometry parameters are listed in Table 3.

Table 3. Mass spectrometry parameters for the detection of ginsenosides.

Compound	Q1 mass	Q3 mass	Collision Energy (V)
Rb3	969.7	789.7	46
Rc	1101.7	335.0	65
Rd	969.8	789.4	60
Re	1101.7	335.0	65
Rg1	823.5	643.5	50

3.7. Statistical Analysis

A p-value < 0.05 was considered statistically significant using t-test for comparing unpaired two means or analysis of variance (ANOVA) with post-hoc Tukey's HSD test for comparing unpaired three means. All data are rounded to three significant digits and expressed as mean ± standard deviation.

4. Conclusions

The present study demonstrated that antioxidant and anti-elastase activities tended to increase as berries ripened from the first week to the last week of July in both three- and four-year-old ginseng berry extracts, and the contents of ginsenosides except Rg1 tended to be the highest in four-year-old ginseng berries harvested in early July. These findings indicate that biological activities and ginsenoside profiles of ginseng berry extracts are dependent on their age and harvest time in July, suggesting the importance of harvest time in developing functional foods and medicinal products of ginseng berry extracts. To the best of our knowledge, this is the first report on the influence of harvest time on biological activity and ginsenoside contents of ginseng berry extracts.

Author Contributions: Conceptualization, S.-Y.S., D.-H.P., I.-S.Y., and S.-S.C.; Formal analysis, S.-Y.S., D.-H.P., S.-W.S., K.-M.P., C.-S.B., H.-S.S., E.I., S.H.R., I.-S.Y., and S.-S.C.; Funding acquisition, I.-S.Y. and S.-S.C.; Investigation, S.-Y.S., D.-H.P., S.-W.S., H.-G.K., J.-H.L., G.Y., J.-H.S., E.I., S.H.R., I.-S.Y., and S.-S.C.; Methodology, S.-Y.S., D.-H.P., S.-W.S., K.-M.P., C.-S.B., H.-S.S., H.-G.K., J.-H.L., G.Y., J.-H.S., I.-S.Y., and S.-S.C.; Writing—original draft, S.-Y.S., D.-H.P., I.-S.Y., and S.-S.C.; Writing—review and editing, S.-W.S., K.-M.P., C.-S.B., H.-S.S., H.-G.K., J.-H.L., G.Y., J.-H.S., E.I., S.H.R., I.-S.Y., and S.-S.C.

Funding: This work was supported by the National Research Foundation of Korea (NRF) grant funded by the Korean government (MSIP; Ministry of Science, ICT & Future Planning) (grant No. NRF-2017R1C1B5015187).

Conflicts of Interest: The authors have declared that there is no conflicts of interest.

References

1. Yun, T.K. Panax ginseng—A non-organ-specific cancer preventive? *Lancet Oncol.* **2001**, *2*, 49–55. [CrossRef]
2. Kim, J.; Cho, S.Y.; Kim, S.H.; Kim, S.; Park, C.-W.; Cho, D.; Seo, D.B.; Shin, S.S. Ginseng berry and its biological effects as a natural phytochemical. *Nat. Prod. Chem. Res.* **2016**, *4*, 209. [CrossRef]
3. Nam, Y.; Bae, J.; Jeong, J.H.; Ko, S.K.; Sohn, U.D. Protective effect of ultrasonication-processed ginseng berry extract on the D-galactosamine/lipopolysaccharide-induced liver injury model in rats. *J. Ginseng Res.* **2018**, *42*, 540–548. [CrossRef] [PubMed]
4. Kim, M.; Yi, Y.S.; Kim, J.; Han, S.Y.; Kim, S.H.; Seo, D.B.; Cho, J.Y.; Shin, S.S. Effect of polysaccharides from a Korean ginseng berry on the immunosenescence of aged mice. *J. Ginseng Res.* **2018**, *42*, 447–454. [CrossRef] [PubMed]
5. Dey, L.; Xie, J.T.; Wang, A.; Wu, J.; Maleckar, S.A.; Yuan, C.S. Anti-hyperglycemic effects of ginseng: Comparison between root and berry. *Phytomedicine* **2003**, *10*, 600–605. [CrossRef] [PubMed]
6. Kim, Y.K.; Yoo, D.S.; Xu, H.; Park, N.I.; Kim, H.H.; Choi, J.E.; Park, S.U. Ginsenoside content of berries and roots of three typical Korean ginseng (*Panax ginseng*) cultivars. *Nat. Prod. Commun.* **2009**, *4*, 903–906. [CrossRef] [PubMed]
7. Han, M.; Fang, X.L. Difference in oral absorption of ginsenoside Rg1 between in vitro and in vivo models. *Acta Pharmacol. Sin.* **2006**, *27*, 499–505. [CrossRef]
8. Han, M.; Sha, X.; Wu, Y.; Fang, X. Oral absorption of ginsenoside Rb1 using in vitro and in vivo models. *Planta Med.* **2006**, *72*, 398–404. [CrossRef]
9. Joo, K.M.; Lee, J.H.; Jeon, H.Y.; Park, C.W.; Hong, D.K.; Jeong, H.J.; Lee, S.J.; Lee, S.Y.; Lim, K.M. Pharmacokinetic study of ginsenoside Re with pure ginsenoside Re and ginseng berry extracts in mouse using ultra performance liquid chromatography/mass spectrometric method. *J. Pharm. Biomed. Anal.* **2010**, *51*, 278–283. [CrossRef]
10. Attele, A.S.; Zhou, Y.P.; Xie, J.T.; Wu, J.A.; Zhang, L.; Dey, L.; Pugh, W.; Rue, P.A.; Polonsky, K.S.; Yuan, C.S. Antidiabetic effects of *Panax ginseng* berry extract and the identification of an effective component. *Diabetes* **2002**, *51*, 1851–1858. [CrossRef]
11. Xie, J.T.; Wu, J.A.; Mehendale, S.; Aung, H.H.; Yuan, C.S. Anti-hyperglycemic effect of the polysaccharides fraction from American ginseng berry extract in ob/ob mice. *Phytomedicine* **2004**, *11*, 182–187. [CrossRef] [PubMed]
12. Park, C.H.; Park, S.K.; Seung, T.W.; Jin, D.E.; Guo, T.; Heo, H.J. Effect of ginseng (*Panax ginseng*) berry EtOAc fraction on cognitive impairment in C57BL/6 mice under high-fat diet inducement. *Evid. Based Complement. Alternat. Med.* **2015**, *2015*, 316527. [CrossRef] [PubMed]
13. Cho, K.S.; Park, C.W.; Kim, C.K.; Jeon, H.Y.; Kim, W.G.; Lee, S.J.; Kim, Y.M.; Lee, J.Y.; Choi, Y.D. Effects of Korean ginseng berry extract (GB0710) on penile erection: Evidence from in vitro and in vivo studies. *Asian J. Androl.* **2013**, *15*, 503–507. [CrossRef] [PubMed]
14. Kim, J.; Cho, S.Y.; Kim, S.H.; Cho, D.; Kim, S.; Park, C.W.; Shimizu, T.; Cho, J.Y.; Seo, D.B.; Shin, S.S. Effects of Korean ginseng berry on skin antipigmentation and antiaging via FoxO3a activation. *J. Ginseng Res.* **2017**, *41*, 277–283. [CrossRef] [PubMed]
15. Xie, J.T.; Wang, C.Z.; Zhang, B.; Mehendale, S.R.; Li, X.L.; Sun, S.; Han, A.H.; Du, W.; He, T.C.; Yuan, C.S. In vitro and in vivo anticancer effects of American ginseng berry: Exploring representative compounds. *Biol. Pharm. Bull.* **2009**, *32*, 1552–1558. [CrossRef] [PubMed]
16. Kim, M.H.; Lee, J.; Jung, S.; Kim, J.W.; Shin, J.H.; Lee, H.J. The involvement of ginseng berry extract in blood flow via regulation of blood coagulation in rats fed a high-fat diet. *J. Ginseng Res.* **2017**, *41*, 120–126. [CrossRef] [PubMed]
17. Choi, H.S.; Kim, S.; Kim, M.J.; Kim, M.S.; Kim, J.; Park, C.W.; Seo, D.; Shin, S.S.; Oh, S.W. Efficacy and safety of *Panax ginseng* berry extract on glycemic control: A 12-wk randomized, double-blind, and placebo-controlled clinical trial. *J. Ginseng Res.* **2018**, *42*, 90–97. [CrossRef] [PubMed]
18. Lee, M.Y.; Seo, H.S.; Singh, D.; Lee, S.J.; Lee, C.H. Unraveling dynamic metabolomes underlying different maturation stages of berries harvested from Panax ginseng. *J. Ginseng Res.* **2019**. [CrossRef]

19. Kim, Y.K.; Yang, T.J.; Kim, S.-U.; Park, S.U. Biochemical and molecular analysis of ginsenoside biosynthesis in Panax ginseng during flower and berry development. *J. Korean Soc. Appl. Biol. Chem.* **2012**, *55*, 27–34. [CrossRef]
20. Sharma, O.P.; Bhat, T.K. DPPH antioxidant assay revisited. *Food Chem.* **2009**, *113*, 1202–1205. [CrossRef]
21. Song, S.H.; Ki, S.H.; Park, D.H.; Moon, H.S.; Lee, C.D.; Yoon, I.S.; Cho, S.S. Quantitative analysis, extraction optimization, and biological evaluation of *Cudrania tricuspidata* leaf and fruit Extracts. *Molecules* **2017**, *22*, 1489. [CrossRef] [PubMed]
22. Chiocchio, I.; Mandrone, M.; Sanna, C.; Maxia, A.; Tacchini, M.; Poli, F.J.I.C. Products Screening of a hundred plant extracts as tyrosinase and elastase inhibitors, two enzymatic targets of cosmetic interest. *Ind. Crop. Prod.* **2018**, *122*, 498–505. [CrossRef]

Sample Availability: Not available.

Article

Liquid Phase and Microwave-Assisted Extractions for Multicomponent Phenolic Pattern Determination of Five Romanian *Galium* Species Coupled with Bioassays

Andrei Mocan [1], Alina Diuzheva [2], Sabin Bădărău [3], Cadmiel Moldovan [1], Vasil Andruch [2], Simone Carradori [4], Cristina Campestre [4], Angela Tartaglia [4], Marta De Simone [4], Dan Vodnar [5], Matteo Tiecco [6], Raimondo Germani [6], Gianina Crişan [1] and Marcello Locatelli [4,*]

[1] Department of Pharmaceutical Botany, "Iuliu Haţieganu" University of Medicine and Pharmacy, 400012 Cluj-Napoca, Romania; mocan.andrei@umfcluj.ro (A.M.); moldovan.cadmiel@yahoo.com (C.M.); gcrisan@umfcluj.ro (G.C.)

[2] Department of Analytical Chemistry, Pavol Jozef Šafárik University, SK-04180 Košice, Slovakia; adyuzheva@gmail.com (A.D.); vasil.andruch@gmail.com (V.A.)

[3] Department of Environmental Sciences, Babeş-Bolyai University, 400084 Cluj-Napoca, Romania; alexandru@transsilvanica.net

[4] Department of Pharmacy, University "G. D'Annunzio" of Chieti-Pescara, 66100 Chieti, Italy; simone.carradori@unich.it (S.C.); cristina.campestre@unich.it (C.C.); angela.tartaglia@unich.it (A.T.); marta.desimone@studenti.unich.it (M.D.S.)

[5] Department of Food Science, University of Agricultural Sciences and Veterinary Medicine, 400372 Cluj-Napoca, Romania; dan.vodnar@usamvcluj.ro

[6] Department of Chemistry, Biology and Biotechnology, University of Perugia, 06132 Perugia, Italy; matteotiecco@gmail.com (M.T.); raimondo.germani@unipg.it (R.G.)

* Correspondence: m.locatelli@unich.it; Tel.: +39-0871-3554590; Fax: +39-0871-3554911

Received: 28 February 2019; Accepted: 26 March 2019; Published: 28 March 2019

Abstract: Background: Galium is a plant rich in iridoid glycosides, flavonoids, anthraquinones, and small amounts of essential oils and vitamin C. Recent works showed the antibacterial, antifungal, antiparasitic, and antioxidant activity of this plant genus. **Methods**: For the determination of the multicomponent phenolic pattern, liquid phase microextraction procedures were applied, combined with HPLC-PDA instrument configuration in five Galium species aerial parts (G. verum, G. album, G. rivale, G. pseudoaristatum, and G. purpureum). Dispersive Liquid–Liquid MicroExtraction (DLLME) with NaCl and NAtural Deep Eutectic Solvent (NADES) medium and Ultrasound-Assisted (UA)-DLLME with β-cyclodextrin medium were optimized. **Results**: The optimal DLLME conditions were found to be: 10 mg of the sample, 10% NaCl, 15% NADES or 1% β-cyclodextrin as extraction solvent—400 μL of ethyl acetate as dispersive solvent—300 μL of ethanol, vortex time—30 s, extraction time—1 min, centrifugation at 12000× g for 5 min. **Conclusions**: These results were compared with microwave-assisted extraction procedures. *G. purpureum* and *G. verum* extracts showed the highest total phenolic and flavonoid content, respectively. The most potent extract in terms of antioxidant capacity was obtained from *G. purpureum*, whereas the extract obtained from *G. album* exhibited the strongest inhibitory effect against tyrosinase.

Keywords: dispersive liquid-liquid microextraction; microwave-assisted extraction; natural deep eutectic solvent; β-cyclodextrin; *Galium* species; tyrosinase inhibition

1. Introduction

The use of plants for the treatment of human diseases is a centuries-old tradition, based on phytotherapy research as well as on ethnopharmacological knowledge. Recently, the use of herbal medicines applied for the prevention and/or preservation of health covers a central role in modern medicine related to the fact that these plant-derived materials avoid the classical side effects of synthetic drugs. Additionally, there are benefits of their long-term historic use—safety, accessibility, and efficacy with a wide range of therapeutic actions [1]. *Galium* is a well-known genus with many medicinal representatives that are rich sources of iridoid glycosides [2–4], flavonoids [5], anthraquinones [6], and small amounts of essential oils and vitamin C [7]. Recent studies showed the antibacterial, antifungal, antiparasitic, and antioxidant activities of representatives of this plant genus [7,8].

G. verum L., also known as Lady's Bedstraw, is an herbaceous perennial plant, native to Europe and Asia, and used commonly in many countries' folk medicine for a large variety of treatments. The dried plants' aerial parts were used to stuff mattresses, and the flowers were also used to coagulate milk for cheese production [9]. The cut and dried aerial parts of the plant, '*Herba gallii verii*', are used for homeopathic purposes. These are still used for exogenous treatment of psoriasis or as a tea with diuretic effect for the cure of pyelitis or cystitis [10]. Moreover, *G. verum* L. has been used as a diuretic for bladder and kidney irritation, externally for poorly healing wounds, as well as for epilepsy and hysteria in Montenegro's traditional medicine [11]. Regarding Turkish folk medicine, it has been used for its diuretic, choleretic, antidiarrheal, and sedative effects [4]. In Romania, the plant is used in traditional medicine mainly for its diuretic, depurative, laxative, sedative, and antirheumatic effects. Additionally, in the Romanian traditional medicine, several *Galium* species are used as components of different cosmetic formulations [12]. *G. album* Mill., the "white bedstraw" or "hedge bedstraw", is an herbaceous annual plant, cited in traditional Albanian pharmacopoeias and folk medicine for healing wounds and gingival inflammations [13]. *G. rivale*, *G. pseudoaristatum*, and *G. purpureum* (syn. *Asperula purpurea*) are less-known species, and to the extent of our knowledge, they have not been investigated yet in terms of chemical composition and antioxidant capacity, nor in terms of enzyme inhibitory potential.

Generally, the use of different extraction procedures on plant-derived material yields different biological activities. In this field, the availability of an efficient, fast, exhaustive, and reproducible extraction procedure allows obtaining a standardized starting material for food additives, nutraceuticals, and phytoformulations. For the extraction of bioactive compounds from *Galium* maceration in methanol [7] or ethanol [14], percolation in methanol [8], and ultrasound-assisted extraction [12] were applied, wherein the extraction time was varied from 30 min to one week. In order to reduce the extraction time and retain or increase the extraction efficiency, new extraction methods are required.

Liquid phase microextraction techniques are positioned as 'green' chemistry methodologies, which require small amounts of organic solvents. In order to make the procedures more environmentally-friendly, ionic liquids (ILs) or natural deep eutectic solvents (NADESs) are frequently used. Comparing ILs with NADESs, more advantages are on the side of NADESs due to their natural original (the main components can be sugars and organic acids), which may vary depending on analysis purpose, making them nontoxic, biodegradable, and incombustible. In comparison with NADESs, most ILs are toxic, have low biodegradability, and have high cost. Either IL or NADES can have high viscosity, so their extracts are limited for direct analysis using HPLC or GC systems [15–18].

Regarding biological activities, in the current work, a key enzyme was considered in order to further evaluate the extracts. Particularly, pigmentation is one of the most obvious phenotypical characteristics in the natural world. Between the pigments, melanin is one of the most widely distributed and is found in bacteria, fungi, plants, and animals. Melanins are heterogeneous polyphenol-like biopolymers with complex structure and color varying from yellow to black. The synthesis of melanin plays an important role in skin color and pigmentation. Tyrosinase, a copper-containing mono-oxygenase, is a key enzyme in melanin biosynthesis [19]. Skin disorders, such as melasma (facial pigmentation), scarce, and freckles, are related to excessive melanin biosynthesis. Thus, tyrosinase inhibitors are used to control or treat pigmentation disorders and are widely used in the cosmetic industry. In fact, some tyrosinase inhibitors, such as kojic acid and hydroquinones, are nowadays commercially produced, but they can present severe side effects, such as skin inflammation [20]. Hence, in recent years, more attention has been paid to the use of natural plant extracts as a safe and alternative source of tyrosinase inhibitors for cosmetic purposes.

In the present study, following our research on innovative microextraction procedures [21–27], different microextraction procedures were examined for the analysis of phenolic compounds in *G. verum* aerial parts, and then applied for the determination of the phenolic pattern of four other *Galium* species (*G. album*, *G. rivale*, *G. pseudoaristatum*, and *G. purpureum*). As an alternative, the microwave-assisted extraction (MAE) technique was used as a reference method [27–29]. To the best of our knowledge, it is the first time that microextraction techniques have been applied for the recovery and the establishment of phenolic compounds in *Galium* species.

2. Results and Discussion

2.1. Preliminary Examinations

Several liquid phase extraction techniques, such as DLLME, UA-DLLME in water, 10% NaCl, NADES, and 1% β-CD media, SA-LLE, and SULLE, were performed in order to select the procedure providing the better quali-quantitative multicomponent profile of phenolic compounds. The extractions were carried out as described in the experimental section. Figure 1 shows that the best results were achieved in the case of DLLME in 10% NaCl and 10% NADES media and UA-DLLME in 1% β-CD. In UA-DLLME, the phase separation was observed only with β-CD, whereas no phase separation was observed using the other additives. The notable increasing of extraction recovery using UA-DLLME with β-CD could be explained because β-CD was able to better dissolve the metabolites in the extraction solvent, contributing to an increased inclusion in its cavity of a higher amount of phenolic compounds. Therefore, DLLME in NaCl and NADES, UA-DLLME in β-CD media were selected for optimization.

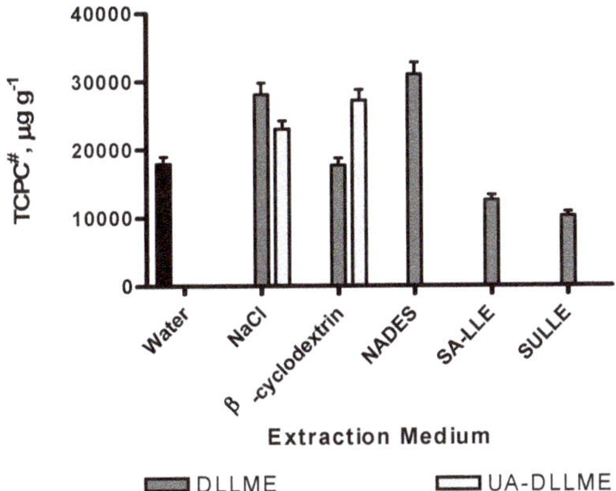

Figure 1. Selection of microextraction procedure. # TCPC—Total concentration of phenolic compounds. Values expressed are means ± S.D. of three measurements. All the values were statistically significant ($p < 0.001$). Raw data regarding the statistical analyses were reported in *Supplementary Materials* section S1.

2.2. Optimization of DLLME and UA-DLLME

Several parameters that could influence the extraction efficiency, such as solid:liquid ratio, extraction and dispersive solvent types, and volume, were selected for optimization. For UA-DLLME, ultrasonication time was also optimized.

2.2.1. Optimization of Extraction Medium Concentration

The extraction medium can significantly affect the extraction yields; therefore, a series of experiments were carried out by adding 5–15% NaCl or NADES solution, or 0.5–1.5% β-CD solution into the vessel containing 10 mg of the dry herbal material. For β-CD, the concentration was lower due to their low water solubility. With 10% NaCl, 15% NADES, and 1% β-CD, the best extraction recoveries were achieved (Figure 2a). Thus, these conditions were applied in further experiments.

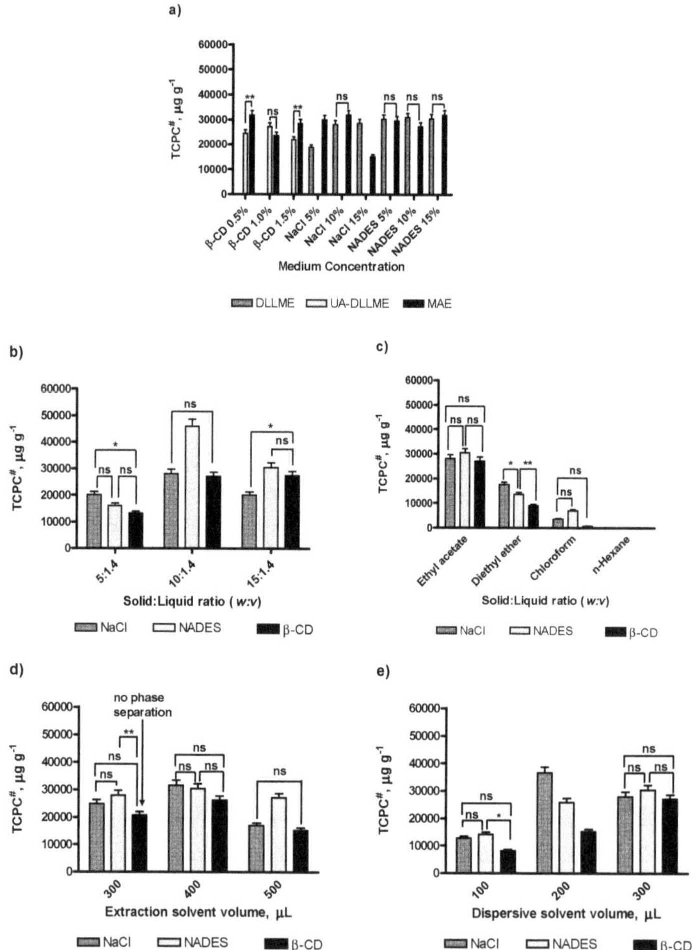

Figure 2. Optimization of DLLME, UA-DLLME, and MAE. (**a**) Effect of medium concentration; (**b**) Effect of solid:liquid ratio; (**c**) Selection of extraction solvent; (**d**) Effect of extraction solvent volume; (**e**) Effect of dispersive solvent volume. # TCPC—Total concentration of phenolic compounds. Values expressed are means ± S.D. of three measurements. All the values were statistically significant ($p < 0.001$), unless otherwise indicated as n.s. (not statistically significant), ** (statistically significant at $p < 0.01$), or * (statistically significant at $p < 0.05$). Raw data regarding the statistical analyses were reported in *Supplementary Materials* section S1.

2.2.2. Optimization of Solid:Liquid Ratio

Three solid:liquid ratios, expressed as mg/mL (5:1.4, 10:1.4, 15:1.4 *w:v*), were examined for their impact on the extraction efficiency. The experimental results showed that the tendency for NaCl and NADES was similar, and the maximum of the extraction recovery was reached with the ratio 10:1.4. For β-CD, with the ratios 10:1.4 and 15:1.4 (*w:v*), no significant differences were observed. Therefore, the optimal solid:liquid ratio was established as 10:1.4 (*w:v*) (Figure 2b). In fact, in the analytical chemistry workflow, if two different systems show similar data, the lower ratio is generally used because it can get the same analytical performances using a lower amount of solvents, raw material, chemicals, etc.

2.2.3. Selection of Extraction Solvent Type and Volume

n-Hexane, ethyl acetate, chloroform, and diethyl ether were tested as potential extractants. The experiments revealed that a higher amount of phenolic compounds was extracted using ethyl acetate (Figure 2c). This could be explained by the different polarities of the extraction solvents and by the interaction with polar phenolic compounds. For instance, with *n*-hexane, a nonpolar solvent, the phenolic compounds were poorly extracted. Diethyl ether and chloroform showed similar extraction efficiency with NaCl and NADES additives, whereas the addition of β-CD did not provide an exhaustive extraction. Taking into account the high volatility of diethyl ether, it was easier to work with ethyl acetate. Therefore, ethyl acetate was selected as appropriate solvent for all samples.

To determine the optimal volume of the extraction solvent, 200, 300, 400, and 500 μL were examined. When the volume is less than 300 μL, the phase separation was not achieved in DLLME and UA-DLLME, while phase separation was reached with 300 μL or more of NaCl and NADES. In order to apply this volume amount to other solvent media, the extraction procedure was modified as follows: the extraction solvent was added in two steps, firstly 200 μL were added in order to achieve an emulsion, then an additional 100 μL of ethyl acetate were rapidly injected. The phase separation was achieved after 5 min in the rest. Applying this procedure, no phase separation in UA-DLLME in β-CD media was observed; therefore, the UA-DLLME was carried out with 400 and 500 μL of ethyl acetate. It was found that with the increase of volume of ethyl acetate, the extraction of total content of phenolics decreased. Therefore, 400 μL was selected for further study on solid samples (Figure 2d).

2.2.4. Selection of Dispersive Solvent Volume

Commonly, ethanol, methanol, and acetonitrile are reported as dispersive solvents in DLLME. In this study, ethanol was selected as dispersive solvent because some food supplements, not considered in this study, of *Galium* are in ethanolic solution. Therefore, the effect of its volume (100–500 μL) on the extraction yields was tested. The results showed that the extraction efficiency was enhanced with the increase of the ethanol volume in the solution until 300 μL, while for higher volumes, phase separation was not achieved (Figure 2e).

2.2.5. Optimization of Ultrasonication Time in UA-DLLME

The cyclodextrins (α, β, γ) show amphiphilic characteristics related to a hydrophilic shell and a hydrophobic cavity and could be usefully used as emulsifiers in order to enhance the extraction recovery for the target analytes. Their capacity to improve the extraction efficiency is related to their ability to reduce the interfacial tension between the two phases by an organic solvent/cyclodextrin complex located in the liquid–liquid interface. In this way, an increased contact area between the two phases was observed [30–33]. The aid of ultrasonication was generally required in order to enhance the solubility, as discussed by Saokham et al. [34] in a recent review paper. Different times have been investigated in the range of 2 to 10 min. Since 5 and 10 min showed similar responses, 5 min was selected as optimal in order to reduce the time of analysis.

2.3. Reference Method: Microwave-Assisted Extraction

In order to evaluate the performances of the proposed procedure, as in the comparison method, MAE was selected and carried out in the same media as LPME procedures at different concentration levels of the solvents (5–15% solution of NaCl and NADES, and 0.5–1.5% solution of β-CD). Figure 2a shows that the extraction efficiencies obtained in 10% NaCl and 15% NADES were comparable to DLLME and UA-DLLME. Therefore, the recovery of total phenolics, using LPME and MAE, was also comparable.

Following our experimental data, the optimized DLLME conditions found were: 10 mg of the sample, 10% NaCl, 15% NADES or 1% β-cyclodextrin, extraction solvent—400 μL of ethyl acetate, dispersive solvent—300 μL of ethanol, vortex time—30 s, extraction time—1 min, centrifugation at 12000× g for 5 min. In the case of UA-DLLME, 5 min of ultrasonication was required.

2.4. Total Phenolic and Flavonoid Content, Antioxidant Capacity, and Tyrosinase Inhibitory Activity

2.4.1. Total Phenolic Content (TPC) by Spectrophotometric Assay

The Folin–Ciocâlteau assay was employed to determine the TPC of *Galium* extracts. The maximum TPC was registered in the ethanolic extract of *G. purpureum* (10.3 ± 0.8 mg GAE/g extract), whereas the lowest concentration was present in the ethanolic extract of *G. rivale* (1.3 ± 0.2 mg GAE/g extract). A recent study by Lakić et al. showed similar results regarding the low phenolic content of *G. verum* (2.4–5.2 mg GAE/g extract), using different extraction solvents [7].

2.4.2. Total Flavonoid Content (TFC) by Spectrophotometric Assay

Results of the total flavonoid content (TFC) of the different plant materials are presented in Table 1. The highest amount for the TFC was obtained for *G. verum* extract, with a value of 8.60 ± 0.07 mg QE/g d.w., comparable with the value obtained for *G. purpureum* extract, containing 8.50 ± 0.04 mg QE/g (d.w.). According to the results of the present study, Vlase et al. reported a TPC of 5.2 ± 0.2 g/100 g for a *G. verum* extract [12] and, additionally, Lakić et al. reported values of 6.4–17.9 mg QE/g (d.w.), for *G. verum* extracts, using different solvents and extraction times [7] confirming the results herein presented.

2.4.3. Antioxidant Potential Assays

The ferric reducing antioxidant power (FRAP), scavenging of DPPH, and ABTS free radical assays were used to evaluate the antioxidant capacity of *Galium* species (Table 1). These methods are simple and widely used for the evaluation of antioxidant capacity of herbal extracts/pure compounds. Moreover, the values regarding the total phenolic content (TPC) and total flavonoid content (TFC) are in accordance with antioxidant capacity values of the extracts. In the DPPH assay, *G. purpureum* (6.3 ± 0.7 mg TE/g extract) exhibited a higher DPPH scavenging capacity than any other considered species (0.4–1.9 mg TE/g extract). The ABTS value for *G. purpureum* (16.7 ± 0.8 mg TE/g extract) was higher in comparison with the values obtained for the other considered species, which ranged from 4.5 to 7.6 mg TE/g extract.

2.4.4. Tyrosinase Inhibitory Activity

Galium extracts had good tyrosinase inhibitory activities (4.66–70.98% at 8 mg/mL), as reported in Table 1. The extract of *G. album* presented the highest tyrosinase inhibition, with a value of 70.98%. Despite the highest concentration of rutin and chlorogenic acid, the ethanolic extract of *G. rivale* showed no inhibitory effect against tyrosinase. This shows that the synergic effect of the compounds from the tested *Galium* samples have no or low inhibitory effects in some cases, although it was demonstrated by many studies that phenolic and flavonoid compounds are, in general, good inhibitors of tyrosinase [19]. The modest tyrosinase inhibitory activity for *Galium* species is confirmed by other studies as well.

For example, Chiocchio et al. reported no tyrosinase inhibitory activity for *G. album* [35]. The low inhibitory activity of these extracts can be explained by the presence of other nondetected compounds, which might block or interfere with the enzyme.

2.5. Quantitative Analysis of Galium Species

The dry extracts were analyzed to establish the fingerprint of phenolic compounds in five *Galium* species. Table 2 summarizes the results obtained by means of a validated HPLC-PDA method for phenolics determination. All measurements were performed in triplicate in order to obtain standard deviation. It can be observed that the amount of the phenolic compounds for all *Galium* species is in the range from 2526.2–11345.1 $\mu g\ g^{-1}$. The major biologically active compounds are chlorogenic acid and rutin. The highest number of the detected phenolic compounds was found in *G. rivale* (11345.1 $\mu g\ g^{-1}$), where the main compound was chlorogenic acid (10192 $\pm$ 34 $\mu g\ g^{-1}$), but the fingerprint was poorer in comparison with other species. The richest multicomponent pattern was observed in *G. pseudoaristatum*, but the quantity of phenolic compounds was the lowest (2526.2 $\mu g\ g^{-1} \pm 46.21$). Chromatograms for each *Galium* sp. were reported in Supplementary Materials section S2.

p-OH benzoic acid, vanillic acid, epicatechin, syringic acid, 3-OH-4-MeO benzaldehyde, *p*-coumaric acid, *t*-ferulic acid, naringin, 2.3-diMeO benzoic acid, benzoic acid, *o*-coumaric acid, harpagoside, *t*-cinnamic acid, and naringenin were not reported into the table because they were not detected by the HPLC-PDA method.

Table 1. Total phenolic and flavonoid content, antioxidant capacity, and enzyme inhibitory effects of the extracts of *Galium* spp. (values expressed are means ± S.D. of three measurements).

	TPC (mg GAE/g Extract)	TFC (mg QE/g Extract)	DPPH Scavenging (mg TE/g Extract)	ABTS Scavenging (mg TE/g Extract)	FRAP (mg TE/g Extract)	Tyrosinase Inhibit. (mg KAE/g Extract)	Tyrosinase Inhibit. (% Inhibition)
G. verum	3.1 ± 0.3 ^{a,e}	8.60 ± 0.07 ^{a,e}	1.9 ± 0.5 ^{a,e}	6.15 ± 0.02 ^{a,e}	21.9 ± 0.9 ^{a,c,e}	7.3 ± 0.2 ^{a,e}	36.96
G. album	2.7 ± 0.1 ^{a,e}	4.88 ± 0.03 ^{a,e}	1.1 ± 0.2 ^{a,e}	6.1 ± 0.1 ^{a,e}	17.19 ± 0.04 ^{b,e}	13.8 ± 3.4 ^{e}	70.98
G. rivale	1.3 ± 0.2 ^{a,d,e}	4.01 ± 0.08 ^{a,c,e}	0.4 ± 0.25 ^{a,e}	4.5 ± 0.5 ^{a,d,e}	12.6 ± 0.2 ^{a,e}	na	na
G. pseudoaristatum	4.5 ± 0.6 ^{a,d,e}	6.7 ± 0.3 ^{a,c}	1.9 ± 0.1 ^{a,e}	7.6 ± 0.2 ^{a,d,e}	19.4 ± 0.9 ^{a,b,c,e}	2.5 ± 1.6 ^{e}	4.66
G. purpureum	10.3 ± 0.8 ^{e}	8.50 ± 0.04 ^{a,e}	6.3 ± 0.7 ^{e}	16.7 ± 0.8 ^{e}	45.2 ± 1.1 ^{e}	6.3 ± 1.4 ^{a,e}	29.71
KA (0.1 mg/mL)							62.52

Na—not active; for tyrosinase inhibition (in percentages), the extracts concentration was 8 mg/mL. TFC = Total Flavonoid Content by spectrophotometric assay; TPC = Total Phenolic Content by spectrophotometric assay; ABTS = 2,2′-azino-bis(3-ethylbenzothiazoline-6-sulphonic acid); DPPH = 2,2-diphenyl-1-picryl-hydrazyl-hydrate; FRAP = Ferric Reducing Antioxidant Power; GAE = Gallic Acid Equivalents; QE = Quercetin Equivalents; TE = Trolox Equivalents; KAE = Kojic Acid Equivalents; KA = Kojic Acid. Raw data regarding the statistical analyses were reported in *Supplementary Materials* section S1. Data marked with different letters indicates significant difference ($p < 0.05$).

Table 2. Phenolic compounds quantified in Galium spp. using DLLME in 15% NADES (values expressed are means ± S.D. of three measurements).

Metabolite	*Galium* Species				
	G. verum	*G. album*	*G. rivale*	*G. pseudoaristatum*	*G. purpureum*
	Conc., µg/g				
	Mean (±S.D.)	Mean (±S.D.)	Mean (±S.D.)	Mean (±S.D.)	Mean (±S.D.)
Gallic acid				109 (±7) ^{a,b,e}	23.2 (±1.5) ^{a,b,e}
Catechin		380 (±96) ^{a,e}	63.5 (±2.5) ^{a,e}	203 (±30) ^{a,e}	
Chlorogenic acid	2986 (±75) ^{e}	8310 (±231) ^{e}	10192 (±34) ^{e}	1640 (±30) ^{e}	5572 (±205) ^{e}
3-OH benzoic acid	853 (±184) ^{e}			87.4 (±12.6) ^{a,c}	374 (±16) ^{b,c,d,e}
Rutin	3624 (±97) ^{e}	275 (±38) ^{a,c,e}	987 (±24) ^{e}	283 (±5) ^{a,b,c,d,e}	137 (±13) ^{a,d,e}
Sinapinic acid				55.7 (±0.9) ^{a,d,e}	203 (±81) ^{a,b}
Quercetin	89.6 (±15.5) ^{a,e}	84.1 (±8.3) ^{a,c,e}		67.9 (±11.6) ^{a,c,e}	
Carvacrol	101 (±2) ^{a,e}	84.1 (±0.9) ^{a,c,e}	102.6 (±0.9) ^{a,e}	81.8 (±2.1) ^{a,c,e}	162 (±26) ^{a,c,e}
Total µg/g	**7654 (±222)**	**9133 (±253)**	**11345 (±42)**	**2526 (±47)**	**6471 (±223)**

Data marked with different letters indicate significant difference ($p < 0.05$).

3. Materials and Methods

3.1. Materials

Chemical standards of phenolic compounds (benzoic acid, carvacrol, catechin, chlorogenic acid, *t*-cinnamic acid, 8-cinnamoyl harpagide (harpagoside), *o*-coumaric acid, *p*-coumaric acid, 2,3-dimethoxybenzoic acid, epicatechin, *t*-ferulic acid, gallic acid, 3-hydroxybenzoic acid, 4-hydroxybenzoic acid, 3-hydroxy-4-methoxybenzaldehyde, naringin, naringenin, quercetin, rutin, sinapinic acid, syringic acid, vanillic acid (all purity > 98%)), β-cyclodextrin (≥97%), *n*-hexane (HPLC-grade), diethyl ether (≥99%), and chloroform (HPLC-grade) were purchased from Sigma-Aldrich (Milan, Italy).

Ethyl acetate (≥99%), acetonitrile (HPLC-grade), methanol (HPLC-grade), ethanol (HPLC-grade), acetic acid (≥99%) as well as D-(+)-glucose were obtained from Carlo Erba Reagents (Milan, Italy). Sodium chloride (≥99%) was obtained from Honeywell (Seelze, Germany). NADES (glycolic acid/betaine mixture) was newly synthesized and supplied by University of Perugia. It was chosen between differently structured novel DES and NADES mixtures for its suitable properties (low freezing point and low viscosity, absence of aromatic compounds in its composition, low cost and natural source of the molecules forming it). Ultra-pure water was obtained using a Millipore Milli-Q Plus water treatment system (18 MΩ cm at 23 °C, Millipore Bedford Corp., Bedford, MA, USA).

3.2. Sampling and Sample Preparation

Samples of *Galium* species were collected from different locations from Romania, as follows: *G. verum* L. from Apuseni mountains region, Sartăș, Alba County, Transylvania, Romania in June 2017, *G. album* Mill. from Podeni, Cluj Coutry, Romania and from Rimetea, Alba Coutry, Romania, *G. purpureum* L. and *G. pseudoaristatum* Schur from Băile Herculane, Caraş-Severin Coutry, in August 2014. All species were authenticated by Dr. Sabin Bădărău and Dr. Andrei Mocan, and voucher specimens were deposited at the herbarium of the Department of Pharmaceutical Botany, "Iuliu Haţieganu" University of Medicine and Pharmacy. Fresh herbal material was dried at room temperature until reaching a constant mass. Afterwards, the plant material was ground into a fine powder using a laboratory mill, mixed to obtain homogenous sample, and kept at 4 °C, for further analyses. All assays were carried out three times (three separate samplings) and in triplicate, and the values reported are represented by average and the standard deviation (S.D.).

3.3. Apparatus

3.3.1. HPLC Analysis

The quantitative analysis of phenolic compounds was performed according to the reported method [36]. The chromatographic system consisted of HPLC Waters liquid chromatograph instrument (model 600 solvent pump, 2996 PDA). Mobile phase was directly on-line degassed by using a Biotech 4CH DEGASI Compact (Onsala, Sweden). For separation of phenolic compounds, C18 reversed-phase column (Prodigy ODS(3), 4.6 × 150 mm, 5 µm; Phenomenex, Torrance, CA), thermostated at 30 °C (±1 °C) was used. The collection and analysis of the data were performed by Empower v.2 software (Waters Corporation, Milford, MA, USA). The mobile phase was a mixture of solution A (3% solution of acetic acid in water) and solution B (3% solution of acetic acid in acetonitrile) in a ratio 93:7 and the gradient mode was applied. The total separation was completed in 1 h (the chemical standards chromatograms, retention times and maximum wavelengths are shown in *Supplementary Materials* section S3).

3.3.2. Auxiliary Equipment

As auxiliary equipment for the extraction procedures, centrifuge model NF048 (Nuve, Ankara, Turkey), vortex (VELP Scientifica Srl, Usmate, Italy), and ultrasonic bath (Falc Instruments, Treviglio,

Italy) were used. MAE was performed using an automatic Biotage Initiator™ 2.0 (Uppsala, Sweden) characterized by 2.45 GHz high-frequency microwaves and power range 0–300 W. An IR sensor probe controlled the internal vial temperature.

3.4. *Extraction Procedures*

Extraction optimization was carried out using *G. verum* and after, under optimized conditions, the microextraction procedure was applied for the other four *Galium* species. The following microextraction procedure were investigated: DLLME, ultrasound-assisted dispersive liquid-liquid microextraction (UA-DLLME), Salting-out liquid-liquid extraction (SA-LLE), and Sugaring-out liquid-liquid extraction (SULLE). The general procedure for the extractions reported in the following paragraphs was described in Figure 3.

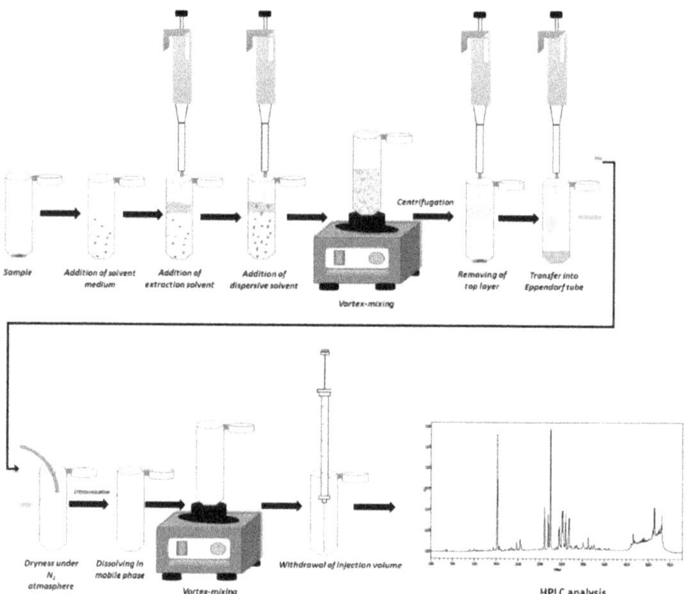

Figure 3. General extraction procedure.

3.4.1. DLLME and UA-DLLME

10 mg of the dry plant material of *G. verum* were accurately weighted and placed into a 2 mL Eppendorf tube. Subsequently, 700 μL of solvent medium (water, 10% NaCl, NADES, IL or 1% β-cyclodextrin (β-CD)), 400 μL of ethyl acetate, and 300 μL of ethanol were added to the Eppendorf tube by automatic pipette. The solution was vortexed during 30 s until a cloudy solution was formed. In the case of UA-DLLME, after those steps, the test tube was placed into the ultrasound bath for 5 min. Then, the solution was kept at rest for 1 min, for the analytes to distribute into the extraction solvent. For the phase separation, the solution was centrifuged at 12000× *g* for 5 min. The extraction solvent was found on the top of the Eppendorf tube, and its whole volume was collected using a microsyringe and transferred to the new Eppendorf tube, and then dried under a gentle stream of nitrogen. The dried residue was redissolved in 50 μL of mobile phase under ultrasonication for 5 min and 20 μL of the obtained solution were injected into the HPLC system.

3.4.2. Salting-Out-LLE

For salting-out-DLLME (SA-LLE), 10 mg of the dry herbal material of *G. verum* were placed into a 2 mL Eppendorf tube. Then, 200 μL of water and 400 μL of acetonitrile were added. To obtain the phase separation, 200 μL of 300 g L^{-1} solution of NaCl were rapidly injected into the Eppendorf tube. The mixture was vortexed for 1 min and a cloudy solution was formed. The next procedures were the same as in the Section 3.4.1.

3.4.3. Sugaring-Out-LLE

For the sugaring-out-LLE (SULLE), the procedure was similar to the SA-LLE. Instead of aqueous NaCl, 200 μL of glucose solution (600 g L^{-1}) was utilized for phase separation.

3.4.4. MAE

10 mg of the dry plant material were placed into a 2 mL sealed vessel suitable for an automatic single-mode microwave reactor and 1 mL of appropriate solvent medium (see Section 3.3.) was added, forming a yellow-green emulsion. MAE was carried out heating by microwave irradiation for 13 min 8 s at 80 °C (which correspond approximatively to 24 h of maceration at 25 °C), and then cooling to room temperature by pressurized air. Then, the homogenate was centrifuged at 12000× *g* for 5 min and 20 μL of the solution were directly injected into the HPLC system.

3.5. Total Phenolic and Flavonoid Content, Antioxidant Capacity, and Tyrosinase Inhibitory Activity

3.5.1. Antioxidant Assays

Total Phenolic Content (TPC) by Spectrophotometric Assay

The TPC was determined using the Folin–Ciocâlteu method described by Mocan et al. [37]. For a high throughput of samples, a SPECTROstar Nano Multi—Detection Microplate Reader with 96-well plates (BMG Labtech, Ortenberg, Germany) was used. Briefly, a mixture solution consisting of 20 μL of extract, 100 μL of Folin-Ciocâlteu reagent, and 80 μL of sodium carbonate (Na_2CO_3, 7.5% *w*/*v*) was homogenized and incubated at room temperature in the dark for 30 min. Afterwards, the absorbance of the samples was measured at 760 nm. Gallic acid was used as a reference standard, and the TPC was expressed as gallic acid equivalents (GAE) in mg/g dry weight (d.w.) of plant material.

Total Flavonoid Content (TFC) by Spectrophotometric Assay

The total flavonoid content (TFC) was calculated and expressed as quercetin equivalents using a method previously described by Mocan et al. [38]. Briefly, a 100 μL aliquot of 2% $AlCl_3$ aqueous solution was mixed with 100 μL of sample. After an incubation time of 15 min, the absorbance of the sample was measured at 420 nm. Quercetin was used as a reference standard, and the TFC was expressed as quercetin equivalents (QE) in mg/g dry weight (d.w.) of plant material.

DPPH Radical Scavenging Assay

The capacity to scavenge the "stable" free radical DPPH, monitored according to the method described by Martins et al. [39], with some modifications, was performed by using a SPECTROstar Nano microplate reader (BMG Labtech, Offenburg, Germany). The reaction mixture in each of the 96-wells consisted of 30 μL of sample solution (in an appropriated dilution) and a 0.004% methanolic solution of DPPH. The mixture was further incubated for 30 min in the dark, and the reduction of the DPPH radical was determined at 515 nm. Trolox was used as a standard reference and the results were expressed as Trolox equivalents per g of dry weight herbal extract (mg TE/g d.w. of herbal extract).

Trolox Equivalent Antioxidant Capacity (TEAC) Assay

In the Trolox equivalent antioxidant capacity (TEAC) assay, the antioxidant capacity is reflected in the ability of the *Galium* extracts to decrease the color, reacting directly with the ABTS radical. The latter was obtained by oxidation of ABTS (2,2'-azinobis(3-ethylbenzothiazoline-6-sulfonic acid)) with potassium peroxydisulfate ($K_2S_2O_8$). The amount of ABTS radical consumed by the tested compound was measured at 760 nm after 6 min of reaction time. The evaluation of the antioxidant capacity was obtained using the total change in absorbance at this wavelength. The percentage of ABTS consumption was transformed in Trolox equivalents (TE) using a calibration curve.

FRAP Assay

In FRAP assays, the reduction of Fe^{3+}-TPTZ to blue-colored Fe^{2+}-TPTZ complex was monitored by the method described by Damiano et al., (2017) with slight modifications [40]. The FRAP reagent was prepared by mixing ten volumes of acetate buffer (300 mM, pH 3.6), one volume of TPTZ solution (10 mM TPTZ in 40 mM HCl) and one volume of $FeCl_3$ solution (20 mM $FeCl_3 \cdot 6H_2O$ in 40 mM HCl). Reaction mixture (25 µL sample and 175 µL FRAP reagent) was incubated in the dark for 30 min at room temperature and the absorbance of each solution was measured at 593 nm using a SPECTROstar Nano Multi-Detection Microplate Reader with 96-well plates (BMG Labtech, Ortenberg, Germany). A Trolox^{TM} calibration curve (0.01–0.10 mg/mL) was plotted as a function of blue-colored Fe^{2+}-TPTZ complex formation, and the results were expressed as milligrams of trolox equivalents (TE) per milligram of extract (mg TE/mg extract).

3.5.2. Tyrosinase Inhibitory Activity

Tyrosinase inhibitory activity of each sample was determined by the method previously described by Likhitwitayawuid and Sritularak, (2001) and Masuda *et al.*, (2005) [41,42] using a SPECTROstar Nano Multi-Detection Microplate Reader with 96-well plates (BMG Labtech, Ortenberg, Germany). Samples were dissolved in water containing 5% DMSO; for each sample four wells were designated as A, B, C, D; each one contained the reaction mixture (200 µL) as follows: (A) 120 µL of 0.66 M phosphate buffer solution (pH = 6.8) (PBS) and 40 µL of mushroom tyrosinase in PBS (46 U/mL) (Tyr), (B) 160 µL of PBS, (C) 80 µL of PBS, 40 µL of Tyr, and 40 µL of sample, and (D) 120 µL of PBS and 40 µL of sample. The plate was then incubated at room temperature for 10 min; after incubation, 40 µL of 2.5 mM L-DOPA in PBS solution were added in each well and the mixtures were incubated again at room temperature for 20 min. The absorbance of each well was measured at 475 nm, and the inhibition percentage of the tyrosinase activity was calculated by the following equation, using as positive control a kojic acid solution (0.10 mg/mL):

$$\%I = \frac{(A-B)-(C-D)}{(A-B)} \times 100 \quad (1)$$

The results were also expressed as mg kojic acid equivalents per gram of dry weight extract (mg KAE/g extract) using a calibration curve between 0.01–0.10 mg kojic acid per milliliter of solution.

3.6. Statistical Analysis

All experiments were performed in triplicate and the results were expressed as the mean value ± standard deviation (S.D.). All comparisons were determined by using two-way ANOVA followed by Bonferroni post-test and GraphPad Prism v.4 for data elaboration. Raw data regarding the statistical analyses were reported in *Supplementary Materials* section S1.

4. Conclusions

In this work, a microextraction procedure was developed and applied for the establishment of the multicomponent phenolic pattern of aerial parts from *G. verum*, *G. album*, *G. rivale*, *G. pseudoaristatum*, and *G. purpureum*. The DLLME procedure in NADES solvent medium could provide high extraction efficiency within a short extraction time and with good correspondence with the MAE procedure. The biological results showed that *G. purpureum* and *G. verum* extracts contained the highest total phenolic and flavonoid contents, respectively. *G. purpureum* extract was the most active extract in terms of antioxidant capacity, whereas the *G. album* extract exhibited the strongest inhibitory effect against tyrosinase, an enzyme involved in several skin disorders. The results indicate that *Galium* extracts have the potential to be used as an alternative source of multifunctional agents and are a promising starting point for development of new bioactive formulations. Further studies are essential for the isolation of pure bioactive compounds and investigation of their molecular mechanisms of action.

Supplementary Materials: The following are available online. In section S1 were reported all the raw data obtained using GraphPad v.4 and Bonferroni post-test related to the statistical analyses for Figures and Tables present in the main text. In section S2 the chromatograms of dry extracts of Galium species obtained using DLLME (@ 278 nm as example of wavelength in which all compounds show absorbance) were reported. In section S3 the chromatogram (@ 278 nm as example of wavelength in which all compounds show absorbance) for the 22 chemical standards, with a table reporting the retention times and the maximum wavelengths used for the quantitative analyses was reported.

Author Contributions: Data curation, A.T., A.D., and M.D.S.; Investigation, A.T., A.D., and M.D.S.; Methodology, M.L., A.M., S.B., C.M., V.A., S.C., C.C., M.T., R.G., D.V. and G.C.; Project administration, M.L., A.M., S.B., C.M., V.A., S.C., D.V. and G.C.; Supervision, M.L., A.M., S.B., C.M., V.A., S.C., D.V. and G.C.

Funding: This work was financially supported by MIUR ex 60%, University "G. d'Annunzio" of Chieti – Pescara, Chieti, Italy, and the Scientific Grant Agency of the Ministry of Education, Science, Research and Sport of the Slovak Republic (VEGA 1/0010/15). Andrei Mocan was supported by the Competitiveness Operational Programme, Priority Axis 1, Action 114, Project Type "Attracting high level personal from abroad" POC-A1-A1.1.4-E_2015 developed with the support of MCI (Acronym: ProGlyCom, Project ID: POC/ID P_37_637, 2016-2020).

Acknowledgments: A.D. gratefully thanks the University "G. d'Annunzio" of Chieti—Pescara, Dept. of Pharmacy and Erasmus+ program for giving her the opportunity of the research project implementation.

Conflicts of Interest: Authors declare no competing financial interest.

Abbreviations

ABTS	2,2′-azino-bis(3-ethylbenzothiazoline-6-sulphonic acid)
β-CD	β-cyclodextrin
DLLME	Dispersive Liquid–Liquid MicroExtraction
DMSO	Dimethyl sulfoxide
DPPH	2,2-diphenyl-1-picryl-hydrazyl-hydrate
d.w.	dry weight
FRAP	Ferric Reducing Antioxidant Power
GAE	Gallic Acid Equivalents
GC	Gas Chromatography
HPLC	High Performance Liquid Chromatography
HPLC-PDA	High Performance Liquid Chromatography-Photodiode array detector
ILs	ionic liquids
KA	Kojic Acid
KAE	Kojic Acid Equivalents
LPME	Liquid Phase MicroExtraction
MAE	Microwave-Assisted Extraction
NADES	NAtural Deep Eutectic Solvent
QE	Quercetin Equivalents
SA-LLE	Salting-out Liquid-liquid extraction
SULLE	Sugaring-out Liquid-liquid extraction
TCPC	Total Concentration of Phenolic Compounds
TE	Trolox Equivalents
TEAC	Trolox Equivalent Antioxidant Capacity
TFC	Total Flavonoid Content by spectrophotometric assay
TPC	Total Phenolic Content by spectrophotometric assay
UA-DLLME	Ultrasound-Assisted Dispersive Liquid–Liquid MicroExtraction

References

1. Efferth, T. Perspectives for globalized natural medicines. *Chin. J. Nat. Med.* **2011**, *9*, 1–6. [CrossRef]
2. Böjthe-Horváth, K.; Kocsis, A.; Párkány, L.; Simon, K. A new iridoid glycoside from *Galium verum* L. First X-ray analysis of a tricyclic iridoid glycoside. *Tetrahedron Lett.* **1982**, *23*, 965–966. [CrossRef]
3. Mitova, M.I.; Anchev, M.E.; Handjieva, N.V.; Popov, S.S. Iridoid patterns in *Galium* L. and some phylogenetic considerations, Zeitschrift Fur Naturforsch. *J. Biosci.* **2002**, *57*, 226–234. [CrossRef]
4. Demirezer, L.O.; Gürbüz, F.; Güvenalp, Z.; Ströch, K.; Zeeck, A. Iridoids, flavonoids and monoterpene glycosides from *Galium verum* subsp. *Verum*. *Turk J. Chem.* **2006**, *30*, 525–534.
5. Zhao, C.C.; Shao, J.H.; Li, X.; Kang, X.D.; Zhang, Y.W.; Meng, D.L.; Li, N. Flavonoids from *Galium verum* L. *J. Asian Nat. Prod. Res.* **2008**, *10*, 611–615. [CrossRef]
6. Zhao, C.C.; Shao, J.H.; Li, X.; Xu, J.; Wang, J.H. A new anthraquinone from *Galium verum* L. *Nat. Prod. Res.* **2006**, *20*, 981–984. [CrossRef]
7. Lakić, N.S.; Mimica-Dukić, N.M.; Isak, J.M.; Božin, B.N. Antioxidant properties of *Galium verum* L. (Rubiaceae) extracts. *Cent. Eur. J. Biol.* **2010**, *5*, 331–337. [CrossRef]
8. Khalili, M.; Ebrahimzadeh, M.A.; Safdari, Y. Antihaemolytic activity of thirty herbal extracts in mouse red blood cells. *Arh. Hig. Rada Toksikol.* **2014**, *65*, 399–406. [CrossRef]
9. Il'ina, T.V.; Kovaleva, A.M.; Goryachaya, O.V.; Aleksandrov, A.N. Essential oil from *Galium verum* flowers. *Chem. Nat. Compd.* **2009**, *45*, 587–588. [CrossRef]
10. Schmidt, M.; Scholz, C.; Gavril, G.; Otto, C.; Polednik, C.; Roller, J.; Hagen, R. Effect of *Galium verum* aqueous extract on growth, motility and gene expression in drug-sensitive and -resistant laryngeal carcinoma cell lines. *Int. J. Oncol.* **2014**, *44*, 745–760. [CrossRef] [PubMed]
11. Menković, N.; Šavikin, K.; Tasić, S.; Zdunić, G.; Stešević, D.; Milosavljević, S.; Vincek, D. Ethnobotanical study on traditional uses of wild medicinal plants in Prokletije Mountains (Montenegro). *J. Ethnopharmacol.* **2011**, *133*, 97–107. [CrossRef]
12. Vlase, L.; Mocan, A.; Hanganu, D.; Gheldiu, A.; Crişan, G. Comparative study of polyphenolic content, antioxidant and antimicrobial activity of four *Galium* species (Rubiaceae). *Dig. J. Nanomater. Biostruct.* **2014**, *9*, 1085–1094.
13. Pieroni, A.; Quave, C.L. Traditional pharmacopoeias and medicines among Albanians and Italians in southern Italy: A comparison. *J. Ethnopharmacol.* **2005**, *101*, 258–270. [CrossRef]
14. Schmidt, M.; Polednik, C.; Roller, J.; Hagen, R. *Galium verum* aqueous extract strongly inhibits the motility of head and neck cancer cell lines and protects mucosal keratinocytes against toxic DNA damage. *Oncol. Rep.* **2014**, *32*, 1296–1302. [CrossRef] [PubMed]
15. Płotka-Wasylka, J.; Rutkowska, M.; Owczarek, K.; Tobiszewski, M.; Namieśnik, J. Extraction with environmentally friendly solvents. *TrAC.* **2017**, *91*, 12–25. [CrossRef]
16. Craveiro, R.; Aroso, I.; Flammia, V.; Carvalho, T.; Viciosa, M.T.; Dionísio, M.; Barreiros, S.; Reis, R.L.; Duarte, A.R.C.; Paiva, A. Properties and thermal behavior of natural deep eutectic solvents. *J. Mol. Liq.* **2016**, *215*, 534–540. [CrossRef]
17. Khezeli, T.; Daneshfar, A.; Sahraei, R. A green ultrasonic-assisted liquid-liquid microextraction based on deep eutectic solvent for the HPLC-UV determination of ferulic, caffeic and cinnamic acid from olive, almond, sesame and cinnamon oil. *Talanta* **2016**, *150*, 577–585. [CrossRef]
18. Shishov, A.; Bulatov, A.; Locatelli, M.; Carradori, S.; Andruch, V. Application of deep eutectic solvents in analytical chemistry. A review. *Microchem. J.* **2017**, *135*, 33–38. [CrossRef]
19. Kim, Y.J.; Uyama, H. Tyrosinase inhibitors from natural and synthetic sources: Structure, inhibition mechanism and perspective for the future. *Cell. Mol. Life Sci.* **2005**, *62*, 1707–1723. [CrossRef] [PubMed]
20. Loizzo, M.R.; Tundis, R.; Menichini, F. Natural and synthetic tyrosinase inhibitors as antibrowning agents: An update. *Compr. Rev. Food Sci. Food Saf.* **2012**, *11*, 378–398. [CrossRef]
21. Zengin, G.; Menghini, L.; Di Sotto, A.; Mancinelli, R.; Sisto, F.; Carradori, S.; Cesa, S.; Fraschetti, C.; Filippi, A.; Angiolella, L.; et al. Chromatographic analyses, in vitro biological activities and cytotoxicity of *Cannabis sativa* L. essential oil: A multidiscipl.inary study. *Molecules* **2018**, *23*, 3266. [CrossRef] [PubMed]

22. Di Sotto, A.; Di Giacomo, S.; Amatore, D.; Locatelli, M.; Vitalone, A.; Toniolo, C.; Rotino, G.L.; Lo Scalzo, R.; Palamara, A.T.; Marcocci, M.E.; et al. A Polyphenol Rich Extract from *Solanum melongena* L. DR2 Peel Exhibits Antioxidant Properties and Anti-Herpes Simplex Virus Type 1 Activity In Vitro. *Molecules* **2018**, *23*, 2066. [CrossRef]
23. Boutaoui, N.; Zaiter, L.; Benayache, F.; Benayache, S.; Cacciagrano, F.; Cesa, S.; Secci, D.; Carradori, S.; Giusti, A.M.; Campestre, C.; et al. *Atriplex mollis* Desf. Aerial Parts: Extraction Procedures, Secondary Metabolites and Color Analysis. *Molecules* **2018**, *23*, 1962. [CrossRef]
24. Locatelli, M.; Macchione, N.; Ferrante, C.; Chiavaroli, A.; Recinella, L.; Carradori, S.; Zengin, G.; Cesa, S.; Leporini, L.; Leone, S.; et al. Graminex pollen: Phenolic pattern, colorimetric analysis and protective effects in immortalized prostate cells (PC3) and rat prostate challenged with LPS. *Molecules* **2018**, *23*, 1145. [CrossRef] [PubMed]
25. Boutaoui, N.; Zaiter, L.; Benayache, F.; Benayache, S.; Carradori, S.; Cesa, S.; Giusti, A.M.; Campestre, C.; Menghini, L.; Innosa, D.; et al. Qualitative and Quantitative Phytochemical Analysis of Different Extracts from *Thymus algeriensis* Aerial Parts. *Molecules* **2018**, *23*, 463. [CrossRef]
26. Melucci, D.; Locatelli, M.; Locatelli, C.; Zappi, A.; De Laurentiis, F.; Carradori, S.; Campestre, C.; Leporini, L.; Zengin, G.; Picot, C.M.N.; et al. A Comparative Assessment of Biological Effects and Chemical Profile of Italian *Asphodeline lutea* Extracts. *Molecules* **2018**, *23*, 461. [CrossRef] [PubMed]
27. Tartaglia, A.; Locatelli, M.; Kabir, A.; Furton, K.G.; Macerola, D.; Sperandio, E.; Piccolantonio, S.; Ulusoy, H.I.; Maroni, F.; Bruni, P.; et al. Comparison between Exhaustive and Equilibrium Extraction Using Different SPE Sorbents and Sol-Gel Carbowax 20M Coated FPSE Media. *Molecules* **2019**, *24*, 382. [CrossRef] [PubMed]
28. Hemwimon, S.; Pavasant, P.; Shotipruk, A. Microwave-assisted extraction of antioxidative anthraquinones from roots of *Morinda citrifolia*. *Sep. Purif. Technol.* **2007**, *54*, 44–50. [CrossRef]
29. Mollica, A.; Locatelli, M.; Macedonio, G.; Carradori, S.; Sobolev, A.P.; De Salvador, R.F.; Monti, S.M.; Buonanno, M.; Zengin, G.; Angeli, A.; et al. Microwave-assisted extraction, HPLC analysis, and inhibitory effects on carbonic anhydrase I, II, VA, and VII isoforms of 14 blueberry Italian cultivars. *J. Enzyme Inhib. Med. Chem.* **2016**, *6366*, 1–6. [CrossRef]
30. Hashizaki, K.; Kageyama, T.; Inoue, M.; Taguchi, H.; Ueda, H.; Saito, Y. Study on preparation and formation mechanism of *n*-alkanol/water emulsion using α-cyclodextrin. *Chem. Pharm. Bull.* **2007**, *55*, 1620–1625. [CrossRef]
31. Inoue, M.; Hashizaki, K.; Taguchi, H.; Saito, Y. Preparation and characterization of n-alkane/water emulsion stabilized by cyclodextrin. *J. Oleo. Sci.* **2009**, *58*, 85–90. [CrossRef]
32. Duchêne, D.; Bochot, A.; Yu, S.C.; Pépin, C.; Seiller, M. Cyclodextrins and emulsions. *Int. J. Pharm.* **2003**, *266*, 85–90. [CrossRef]
33. Diuzheva, A.; Carradori, S.; Andruch, V.; Locatelli, M.; De Luca, E.; Tiecco, M.; Germani, R.; Menghini, L.; Nocentini, A.; Gratteri, P.; et al. Use of Innovative (Micro)Extraction Techniques to Characterise *Harpagophytum procumbens* Root and its Commercial Food Supplements. *Phytochem. Anal.* **2018**, *3*, 233–241. [CrossRef] [PubMed]
34. Saokham, P.; Muankaew, C.; Jansook, P.; Loftsson, T. Solubility of Cyclodextrins and Drug/Cyclodextrin Complexes. *Molecules* **2018**, *23*, 1161. [CrossRef]
35. Chiocchio, I.; Mandrone, M.; Sanna, C.; Maxia, A.; Tacchini, M.; Poli, F. Industrial Crops & Products screening of a hundred plant extracts as tyrosinase and elastase inhibitors, two enzymatic targets of cosmetic interest. *Ind. Crop. Prod.* **2018**, *122*, 498–505. [CrossRef]
36. Zengin, G.; Menghini, L.; Malatesta, L.; De Luca, E.; Bellagamba, G.; Uysal, S.; Aktumsek, A.; Locatelli, M. Comparative study of biological activities and multicomponent pattern of two wild Turkish species: *Asphodeline anatolica* and *Potentilla speciosa*. *J. Enzyme Inhib. Med. Chem.* **2016**, *31*, 203–208. [CrossRef]
37. Mocan, A.; Schafberg, M.; Crisan, G.; Rohn, S. Determination of lignans and phenolic components of *Schisandra chinensis* (Turcz.) Baill. using HPLC-ESI-ToF-MS and HPLC-online TEAC: Contribution of individual components to overall antioxidant activity and comparison with traditional antioxidant assays. *J. Funct. Food.* **2016**, *24*, 579–594. [CrossRef]
38. Mocan, A.; Crişan, G.; Vlase, L.; Crişan, O.; Vodnar, D.C.; Raita, O.; Gheldiu, A.M.; Toiu, A.; Oprean, R.; Tilea, I. Comparative studies on polyphenolic composition, antioxidant and antimicrobial activities of *Schisandra chinensis* leaves and fruits. *Molecules* **2014**, *19*, 15162–15179. [CrossRef] [PubMed]

39. Martins, N.; Barros, L.; Dueñas, M.; Santos-Buelga, C.; Ferreira, I.C.F.R. Characterization of phenolic compounds and antioxidant properties of *Glycyrrhiza glabra* L. rhizomes and roots. *RSC Adv.* **2015**, *5*, 26991–26997. [CrossRef]
40. Damiano, S.; Forino, M.; De, A.; Vitali, L.A.; Lupidi, G.; Taglialatela-Scafati, O. Antioxidant and antibiofilm activities of secondary metabolites from *Ziziphus jujuba* leaves used for infusion preparation. *Food Chem.* **2017**, *230*, 24–29. [CrossRef] [PubMed]
41. Likhitwitayawuid, K.; Sritularak, B. A new dimeric stilbene with tyrosinase inhibitory activity from *Artocarpus gomezianus*. *J. Nat. Prod.* **2001**, *64*, 1457–1459. [CrossRef] [PubMed]
42. Masuda, T.; Yamashita, D.; Takeda, Y.; Yonemori, S. Screening for tyrosinase inhibitors among extracts of seashore plants and identification of potent inhibitors from *Garcinia subelliptica*. *Biosci. Biotechnol. Biochem.* **2005**, *69*, 197–201. [CrossRef] [PubMed]

Sample Availability: Samples of the compounds are not available from the authors.

Article

Synthesis of Graphene Oxide Based Sponges and Their Study as Sorbents for Sample Preparation of Cow Milk Prior to HPLC Determination of Sulfonamides

Martha Maggira [1], Eleni A. Deliyanni [2] and Victoria F. Samanidou [2,*]

1 Laboratory of Analytical Chemistry, Department of Chemistry, Aristotle University of Thessaloniki, GR-541 24 Thessaloniki, Greece; marthamaggira@gmail.com

2 Laboratory of General and Environmental Technology, Department of Chemistry, Aristotle University of Thessaloniki, GR-541 24 Thessaloniki, Greece; lenadj@chem.auth.gr

* Correspondence: samanidu@chem.auth.gr; Tel.: +30-2310997698; Fax: +30-2310997719

Received: 8 May 2019; Accepted: 31 May 2019; Published: 31 May 2019

Abstract: In the present study, a novel, simple, and fast sample preparation technique is described for the determination of four sulfonamides (SAs), namely Sulfathiazole (STZ), sulfamethizole (SMT), sulfadiazine (SDZ), and sulfanilamide (SN) in cow milk prior to HPLC. This method takes advantage of a novel material that combines the extractive properties of graphene oxide (GO) and the known properties of common polyurethane sponge (PU) and that makes sample preparation easy, fast, cheap and efficient. The PU-GO sponge was prepared by an easy and fast procedure and was characterized with FTIR spectroscopy. After the preparation of the sorbent material, a specific extraction protocol was optimized and combined with HPLC-UV determination could be applied for the sensitive analysis of trace SAs in milk. The proposed method showed good linearity while the coefficients of determination (R^2) were found to be high (0.991–0.998). Accuracy observed was within the range 90.2–112.1% and precision was less than 12.5%. Limit of quantification for all analytes in milk was 50 μg kg^{-1}. Furthermore, the PU-GO sponge as sorbent material offered a very clean extract, since no matrix effect was observed.

Keywords: sulfonamides; HPLC; graphene oxide; sponge; milk

---

1. Introduction

Sulfonamides are a group of synthetic antibacterial agents, which are widely used in veterinary practice for prophylactic and therapeutic purposes and as feed additives. Due to their ability to inhibit folic acid synthesis in microorganisms, they are commonly used against a wide range of bacteria, protozoa, parasites, and fungi [1–3].

However, the improper administration of sulfa drugs in dairy husbandry and the insufficient withdrawal periods can lead to noncompliant residues in animal originated foods, a fact which can contribute to several concerns in the dairy industry and public health [4].

In humans, such concerns comprise the rise of allergic or toxic reactions and the development of drug-resistance, whereas in the dairy industry they provoke the inhibition of bacterial fermentation in cheese and yoghurt production [5]. In order to safeguard public health and ensure food safety, monitoring of such residues in products designated for human consumption is considered mandatory. For this reason, the European Union has established a maximum residue level (MRL) for sulfonamides in foodstuffs of animal origin, which in the case of milk is 100 μg kg^{-1} [6]

Additionally, several methods have been described for the detection and/or determination of sulfonamides in foods of animal origin such as microbial inhibition assays, immunochemical methods, capillary electrophoresis (CE), gas chromatography (GC), and HPLC [5,7].

Sample preparation is a key step prior to the detection of sulfonamides present in different kinds of samples. The clean-up procedure of various matrices can be accomplished by either traditional techniques, such as liquid-liquid extraction (LLE) [8], or modern methods, like solid phase extraction (SPE) [9], solid phase micro extraction (SPME) [1,10], fabric phase solid extraction [11], matrix solid phase dispersion (MSPD) [12] and Quick, Easy, Cheap, Effective, Rugged and Safe (QuEChERS) method [13,14]. Most of the aforementioned techniques depend on an absorbent material to achieve high analytical specificity and selectivity.

However, in the analysis of complex matrices, many innovative materials have emerged as valuable tools to enhance the efficiency of the extraction and isolation of the target analytes. As such, graphene-based materials are preferred to other carbon-based nanomaterials due to their great potential on the sample preparation procedure. Graphene (G) is a two dimensional nanomaterial with extraordinary physicochemical properties such as thermal and chemical stability, thermal conductivity, hydrophobicity, and large specific surface area [15]. Graphene oxide (GO) is a single-atomic layered material, an important derivative of graphene with similar structure, which is composed easily from the oxidation of graphite. However, GO is more polar than G because of the hydroxyl (–OH) and carboxyl (–COOH) groups, a characteristic that facilitates GO bonds into other compounds such as aminopropyl silica [16].

Graphene based materials are extensively applied in SPE procedure as they offer high sorption efficiency for organic compounds and metal ions mainly in environmental samples [17–19]. Although G and GO demonstrate excellent sorbent characteristics, many limitations have been reported concerning their isolation from well dispersed solutions and their sheets' restacking or escaping from the SPE column [20,21].

In order to surpass the problems having occurred during the elution and sample loading in SPE, new sample preparation techniques have been developed such as the use of graphene-based materials in dispersive solid phase extraction (DSPE) and MSPD. In DSPE the absorbent is mainly utilized in food [22] and environmental samples [23–26], whereas MSPD has been performed for the extraction of sulfonamides in milk samples [27].

Recently, melamine sponge was functionalized with graphene, via a microwave-assisted hydrothermal process, in order to be used as adsorbent for SAs extraction from milk, egg, and environmental water [28]. The proposed method was highly accurate and sensitive for the analysis of nine SA's. However, it is not referred to the determination of sulfathiazole (STZ), sulfamethizole (SMT), and sulfanilamide (SN). In the current study, commercial polyurethane (PU) sponges, a kind of cheap porous material, were examined for SAs extraction from milk. PU sponges, compared with other sponge materials, such as melamine [29,30], and chitosan sponge [31] present certain advantages like easy access, low cost, and high resilience, excellent flexibility, and reuse [32]. Moreover, the surface of the PU sponge was used as a skeleton for hydrophobic modifiers. Hence, in the current study, surface modification was achieved via a green route at ambient conditions.

Polyurethane (PU) sponges with a unique 3D structure have a potential application as absorbents due to their advantages of easy access, low cost, and high resilience compared to other porous materials, such as melamine foam and chitosan sponge. Although PU sponge is hydrophilic, modifications or physical coating like functionalization with graphene are required to increase the hydrophobicity and are usually used to achieve higher efficiency in separations [32].

Consequently, the objective of this study was to combine the unique properties of PU sponge being functionalized with GO in order to serve as an innovative absorbent material in the sample preparation procedures. Due to its properties of low cost, time saving, and simplicity, the GO-PU material was further used for the determination of sulfonamides in cow milk samples prior to HPLC-DAD method.

2. Results and Discussion

2.1. Characterization

Polyurethane sponge was used as a base material in order to be functionalized with graphene oxide. Polyurethane presents an open-hole structure, with a high porosity as well as a rich surface chemistry with surface-groups that can attract and react with different molecules. Graphene oxide was embodied in the PU skeleton after the dispersion of GO in water. Graphene oxide was connected to polyurethane after chemical interactions between the GO (epoxy-groups) and polyurethane surface groups (C=O and –N–H groups). After the polyurethane functionalization with graphene oxide, the sponge prepared appeared with a black color and presented hydrophobicity that was further increased after the coating with PVA.

The XRD diffraction patterns of the prepared graphite oxide (GO) as well as of the GO impregnated sponge before (PU-GO) and after the PVA coating (PU-GO-PVA) are presented in Figure 1. Graphite presents a sharp diffraction peak at 26.6° in the XRD pattern (not presented), attributed to interlayer (002) spacing (d = 0.33 nm). The characteristic XRD peak of graphite oxide appeared at $2\theta = 10.9°$; as estimated by the Bragg's law, the interlayer distance between the carbon layers, increased from 0.33 nm for graphite to 0.81 nm for GO [33]. In the XRD pattern of the GO impregnated sponge (PU-GO) the characteristic XRD peak of graphite oxide, at $2\theta = 10.9°$, was not present, indicating that the layered structure of GO was destroyed. A diffraction peak at $2\theta = 21°$ could be due to PVA while the broad peaks at around 11.6° and 19.8° indicated some degree of crystallinity of the PU [34–36]. The XRD pattern for the sample after the sulfonamide adsorption (PU-GO-SA), which is also presented in Figure 2, reveals that a decrease of crystallinity was observed, evidenced by the disappearance of the peak at $2\theta = 11.6°$.

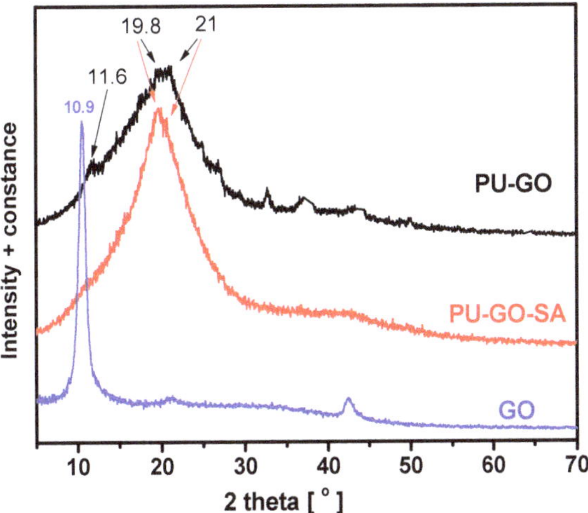

Figure 1. X-ray diffraction (XRD) patterns of the graphite oxide (GO), the graphene oxide impregnated sponge (PU-GO), and the sponge after the adsorption of sulfonamides (PU-GO-SA).

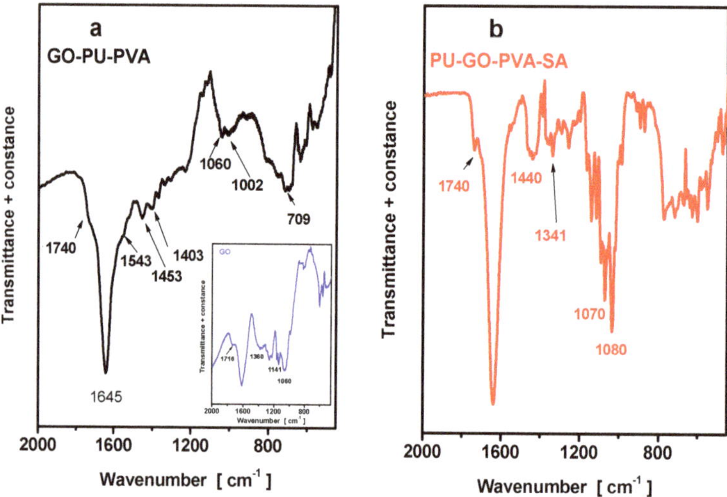

Figure 2. Fourier-transform infrared (FTIR) spectra for (**a**) polyurethane-graphene oxide- polyvinyl alcohol (PU-GO-PVA) sponge raw and after (**b**) the absorption of sulfonamide's (SA's) (PU-GO-PVA-SA)-(in the inset the spectrum of GO).

FTIR spectroscopy was used in this study to identify the possible interactions between GO and PU (PU-GO), between PU-GO and PVA (PU-GO-PVA sponge) as well as between the sponge and the sulfonamides (PU-GO-PVA-SA) in order for the adsorption mechanism to be revealed. The FTIR spectra of PU-GO-PVA as well as of PU-GO-PVA after the sorption of sulfonamide (PU-GO-PVA-SA), are presented in Figure 2. The FTIR spectra of GO is presented in the inset of Figure 2. GO contains polar groups on the edges of graphite layers such as carbonyl, carboxyl, and epoxide, as well as hydroxyl groups within the basal planes of the graphene sheets. In the spectrum of GO (Figure 2a), the bands at 1050–1100 cm^{-1} and ~1716 cm^{-1} can be attributed to carboxylic groups whereas the band at ~1600 cm^{-1} can be attributed to C=C stretching mode of the sp^2 carbon skeletal network and/or to epoxy groups. The band at 1356 cm^{-1} is due to C–OH stretching of O–H groups, while the band at 1045 and at 1141 cm^{-1} can be also attributed to epoxy and alkoxy C–O groups, respectively.

Polyurethane (PU) is a polymer obtained after the polymerization of diisocyanate and polyol that contains C=O and –NH groups (electron donating sites); these groups are able to form hydrogen bonds with graphene oxide during the complexation. The spectra of PU-GO-PVA sponge presented peaks at 1740 and 1060 cm^{-1} attributed to carboxyl and epoxy groups, respectively, at a lower intensity compared to the relative peaks of the spectra of GO, indicating the involvement of these groups in the composite synthesis. The peaks at 1543 cm^{-1} could be attributed to amide II formation after reaction of the carboxylic groups of GO with –NH groups of PU while the peaks at about 1453 cm^{-1} could be attributed to –CH_3 groups of PVA indicating the covering [37–39].

The most significant spectra alterations for the GO-PU-PVA after the SA adsorption (GO-PU-PVA-SA sample), are the new bands appearing at 1260 and 1070 cm^{-1} in addition to the diminishing of the peaks at 1191, 1130 and 1740 cm^{-1} (carbonyl) absorption bands (Figure 2). The new band at 1440 cm^{-1} can be attributed to amide I formation due to interactions between the SA amines and the sponge carboxylates, causing the diminishing of the band at 1740 cm^{-1}. The new band at 1260 cm^{-1}, can be attributed to hydrogen bond interaction between the GO-PU-PVA carboxyl groups and the sulfones/O=S=O groups of SA which are strong hydrogen-bond acceptors. It is obvious that the grafting of PU with extra carboxyl groups enhanced the SA adsorption owning to their reactions with the amines and the hydrogen bond with the sulfones/O=S=O groups of the SA. This was also reported for dorzolamine encapsulation to chitosan, as well as for pramipexole adsorption on activated carbon.

2.2. Synthesis Optimization

The mass of the material retained in the sponge was initially studied, keeping its second mass at 0.04 g. After selecting three different levels (0.12, 0.24, and 0.32 g), the procedure of the sponge preparation was followed. The sample preparation was performed in standard solutions with all three materials. From the results as presented in Figure 3, it seems that the mass of 0.12 g is more effective for the adsorption.

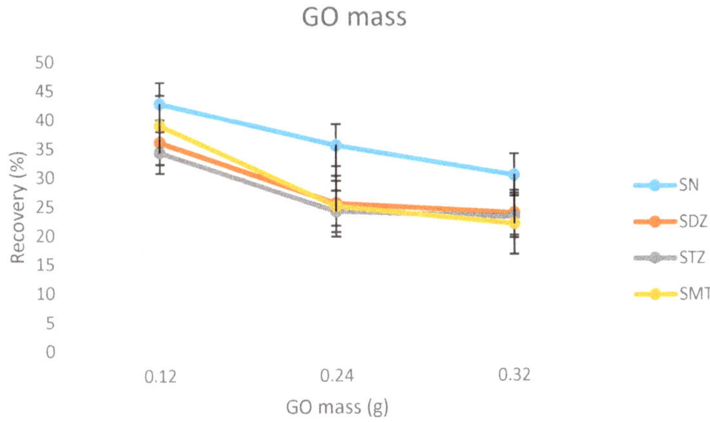

Figure 3. Effect of the graphene oxide (GO) mass on the adsorption efficiency of the sulfonamides.

The size of the sponge was optimized after the testing of two different sizes. Particularly 0.04 g and 0.07 g sponge were dipped in the dispersed solution. The results showed that the bigger sponge is sufficient to achieve the optimum adsorption.

For the PU-GO sponge formation, the GO molecules should be immobilized during its preparation. This is accomplished with the adding of a solvent like water or some polymer, of which polyvinyl alcohol (PVA) is more common due to its low cost. In the present research, two such solvents were tested, water and PVA. As shown in the results (Figure 4), PVA helps in the sample preparation procedure.

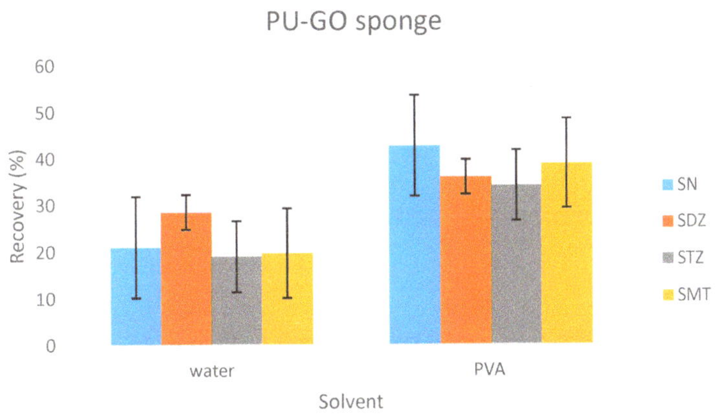

Figure 4. Effect of the solvent in the absolute recoveries of the four sulfonamides.

Different solutions of NH_3/EtOH containing 60 mL of the mixture were prepared in three different volume ratios (4:1, 1:1 and 1:4) and were further applied in the functionalization of the GO-PU material. The results revealed that the quantity of NH_3 was crucial to the absorption and that the volume ratio 4:1 achieved higher efficiency.

2.3. Chromatography

The target analytes were separated by gradient elution. Optimum gradient program was chosen as providing good analytes' resolution, at the shortest analysis. A typical chromatogram is shown in Figure 5. The retention times were observed at 6.345, 7.566, 8.748, and 12.899 min for SN, SDZ, STZ, and SMT respectively.

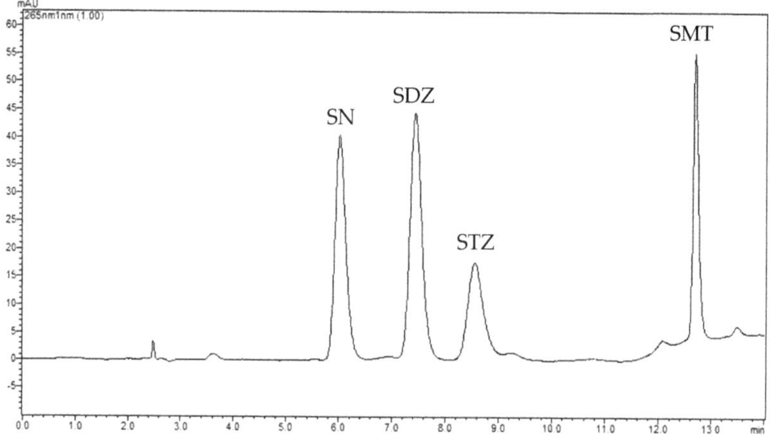

Figure 5. A typical HPLC chromatogram of standard solution of examined analytes at the concentration of 5 ng μL^{-1}. Peaks are as follows: SN: 6.345 min, SDZ: 7.566 min, STZ: 8.748 min, and SMT: 12.899 min.

2.4. Sample Preparation Optimization

All initial optimization experiments were performed using standard solutions of sulfonamides. The optimum conditions established were further checked for their appropriateness to the milk matrix.

In the loading and elution step different methods were tested. Although stirring showed the best results in the tests with the standard solutions, as shown in Table 1 the extraction declined sharply when the milk samples were tested and the recovery rates ranged from 7 to 14%. Thus, centrifugation in low rates was selected. Centrifugation at low rates had two purposes: (1) sufficient sample interaction with the material, and (2) preventing the adsorbent from escaping from the structure of the sponge. High centrifugation rates hindered the extraction process. With regards to sonication, GO particles were released from the sponge and sample handling was difficult.

Table 1. Effect of the loading/elution time and the extraction procedure on the efficiency of the method. (SN = sulfanilamide, SDZ = sulfadiazine, STZ = sulfathiazole, SMT = sulfamethizole)

Loading/Elution Time (min)		Absolute Recovery Rates (R%)			
		SN	SDZ	STZ	SMT
Rest	15/15	21.1	23.5	29.9	29.3
Sonication	7/7	15.6	21.2	29.7	33.7
Stirring	15/15	22.3	28.7	34.8	36.4
Centrifugation	15/15	30.9	24.0	27.6	29.6

Additionally, the volume of the sample, the elution solvents, the size of the sponge, loading and elution time, and the pH were optimized. The extraction was conducted with two different volume samples (1.5 and 3 g) that were spiked with the same amount of the target analytes. The results revealed a decrease in the extraction efficiency by increasing the volume of the sample.

With regards to the elution, methanol (MeOH) and acetonitrile (ACN) were tested both separately and in mixture. It is obvious from the results that the mixed solution increases the efficiency of the elution. In order to succeed better results, 1% acetic acid was added. The addition of acetic acid was successful and the optimum volume ratio for the CH_3COOH/ACN/MeOH solution was 50:40:10.

As for the loading and elution time 10, 15, 20 min were tested. From the results it is observed that 10 min are not enough for the loading and the extraction of the target analytes. However, 15 and 20 min yielded similar results, and the shortest time was selected to reduce the process time.

The effect of the pH in the extraction efficiency was tested, adding 0.5 mL of buffer solution into the sample. Table 2 presents the results obtained from the addition of pH 3, 5, 7, and 9 buffer solution in milk sample. It is obvious from the results that the optimum pH is 5, whereas lower or higher pH values results in decrease in the adsorption for all SAs.

Table 2. Effect of the pH on the adsorption efficiency of the four sulfonamides. (SN = sulfanilamide, SDZ = sulfadiazine, STZ = sulfathiazole, SMT = sulfamethizole). Optimum pH value is given in bold.

	Absolute Recovery Rates (R%)			
pH	**SN**	**SDZ**	**STZ**	**SMT**
3	21.8	22.0	31.7	29.3
5	**22.2**	**27.5**	**36.1**	**31.7**
7	12.3	17.1	21.7	17.2
9	15.0	22.0	27.9	19.0

The proposed sample preparation protocol is very simple and rapid, with low consumption of organic solvents and very clean background signal. Figure 6 illustrates the simple pretreatment procedure. Typical chromatograms of a blank and a spiked milk sample are shown in Figure 7a,b. It is clear that the peaks of the substrate do not interfere with the analysis as they elute at different times.

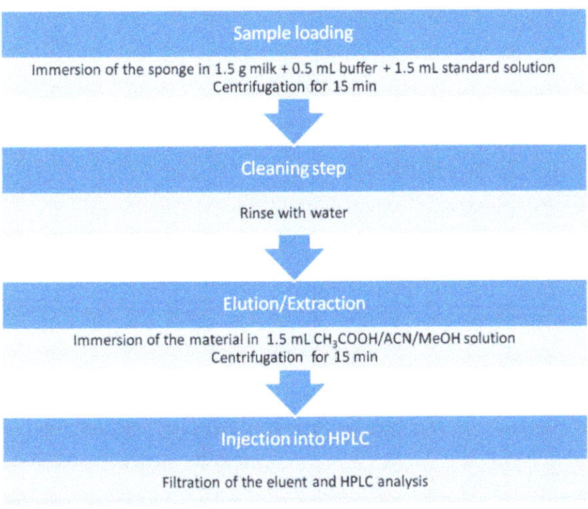

Figure 6. Steps of sample preparation procedure.

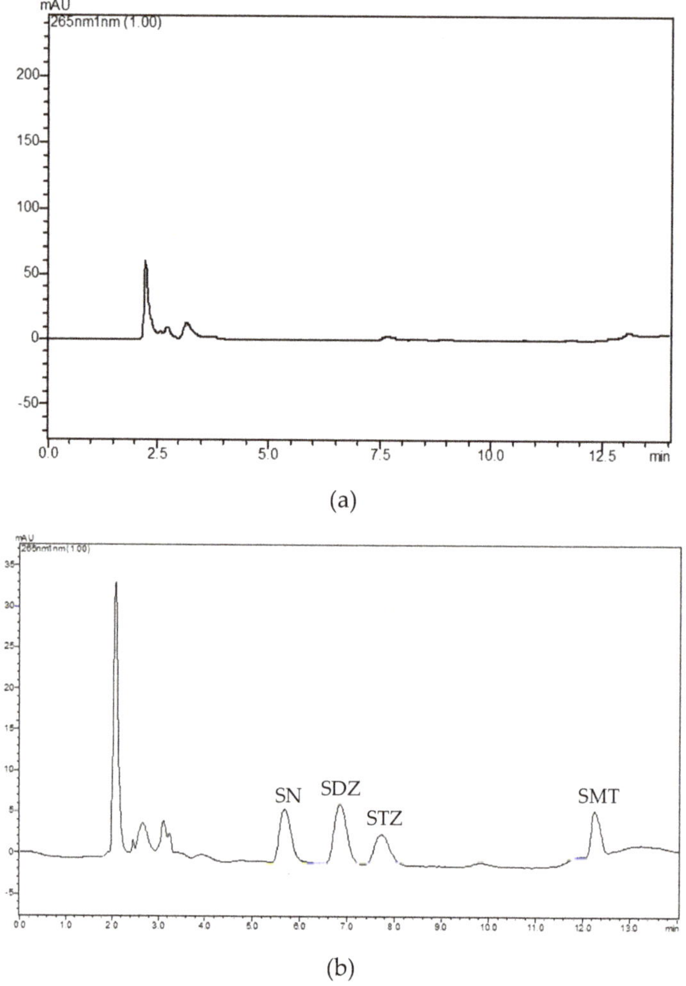

Figure 7. Chromatogram of (**a**) blank milk sample and (**b**) spiked milk sample at a concentration of 300 μg kg^{-1}.

2.5. Method Validation

2.5.1. Selectivity

The good resolution between the chromatographic peaks of analytes and the absence of interferences in the spiked milk samples indicate that a good selectivity was achieved.

2.5.2. Linearity and Sensitivity

Standard solutions showed linearity for all of the target analytes within the range of 0.5 to 10 ng μL^{-1} and showed and good correlation coefficients (0.981–0.999). Moreover, calibration curves were constructed using fortified milk samples after sample preparation, and good coefficients of determination between 0.9969 and 0.999 were achieved over the examined range. (Table 3). Limit of quantification for all analytes in milk was 50 μg kg^{-1}.

Table 3. Linearity data in standard solutions and spiked milk samples. (SN = sulfanilamide, SDZ = sulfadiazine, STZ = sulfathiazole, SMT = sulfamethizole).

Analytes	Calibration Curve	Coefficients of Determination (R^2)
	Standard Solutions	
SN	y = 119367x + 366.66	0.999
SDZ	y = 121371x + 27142	0.995
STZ	y = 64281x + 16245	0.999
SMT	y = 58218x + 13560	0.981
	Milk	
SN	y = 14785x − 1957.4	0.996
SDZ	y = 17012x − 3034.3	0.998
STZ	y = 10182x − 2023.9	0.991
SMT	y = 8489.3x − 2609.2	0.991

2.5.3. Precision and Accuracy

The precision of the method was based on within-day repeatability and between-day precision. The former was assessed by replicate (n = 4) measurements from a spiked milk sample at the MRL level for all examined sulfonamides. The recoveries of spiked samples were calculated by comparison of the peak area ratios for extracted compounds toward the values derived from spiked calibration curves. In Between-day reproducibility a triplicate determination was performed for a period of three days (Table 4). Precision and accuracy was determined at three concentration levels according to the 657/2002/EC decision [40].

Table 4. Precision and accuracy parameters of the method for the determination of sulfonamides in milk samples. (SN = sulfanilamide, SDZ = sulfadiazine, STZ = sulfathiazole, SMT = sulfamethizole).

Added Concentration (μg kg^{-1})	Analyte	Intra-Day n = 4		Inter-Day n = 3 × 3	
		R%	RSD	R%	RSD
50	SN	98.2	7.6	97.6	7.1
	SDZ	106.7	6.9	104.3	3.3
	STZ	93.6	8.5	95.6	9.8
	SMT	93.4	10.4	90.8	11.0
100	SN	103.3	4.0	107.7	4.0
	SDZ	112.1	10.8	105.3	0.4
	STZ	96.8	11.0	90.2	10.9
	SMT	92.8	11.8	97.6	12.4
150	SN	100.2	10.4	102.4	7.6
	SDZ	108.7	3.0	100.1	6.0
	STZ	96.6	9.8	92.8	12.0
	SMT	101.7	10.3	95.3	9.5

2.5.4. Decision Limit and Capability of Detection

Decision limit (CCα) is defined as "the limit at and above which it can be concluded with an error probability" and it was calculated after the analysis of 20 spiked milk samples at the MRLs of each compound. The decision limits CCa were 100.2 μg kg^{-1} for SN, 100.3 μg kg^{-1} for SDZ, 100.4 μg kg^{-1}

for STZ, and μg kg^{-1} for 100.3 SMT. Capability of detection (CCb) defined as "the smallest content of the substance that may be detected, identified, and/or quantified in a sample with an error probability of b" and it was calculated after the spiking of 20 blank milk samples at the CCa level of each compound. The capability of detection (CCb) were 110.7 μg kg^{-1} for SN, 109.3 μg kg^{-1} for SDZ, 115.4 μg kg^{-1} for STZ, and 114.3 μg kg^{-1} for SMT.

2.6. Application to Real Samples

The method was applied for the determination of the examined analytes in cow milk samples from local food stores. Five random samples of three different types of milk were collected and analyzed, including full-fat (3.5%), semi-skimmed (1.5%), and skimmed (0%) milk. All analyzed samples were negative in the presence of examined analytes.

2.7. Comparison with Other Methods

The method described in this study was compared with previous analytical approaches for the determination of SAs in milk. The analysis' results are comparable with those attained by other methods, with fairly good recoveries and quite satisfactory sensitivity. Although it provides higher LODs and LOQs than previously reported methods, it is a less costly (no commercial SPE products are needed) and less time-consuming method with easy handling of sponge and does not require highly sophisticated equipment since no MS is used (Table 5).

Table 5. Performance of the presented method in comparison with previously reported analytical methods.

Analytes	Sample Preparation	Analytical Technique	Run Time (min)	LOD-LOQ	Recovery (%)	Ref
4SAs	MSPE	HPLC-AS	N/A	LOD (ng/mL): 2.0–2.5 LOQ (ng/mL): 6.0–7.5	92–105	[41]
38 veterinary drugs (18SAs)	SPE	UHPLC-ESI-MS/MS	13.5	CCα (μg/kg): 109–114 (SAs) CCβ (μg/kg): 116–123 (SAs)	87–119 (all analytes)	[42]
6SAs	SPE	HPLC-DAD	15.3	LOD (μg/kg): 1.9–13.3 LOQ (μg/kg): 5.6–42.2	N/A	[7]
9 SAs	MSPE	HPLC DAD	35	LOD (μg/L): 7–14	81.8–114.9	[27]
5 SAs	MSPE	HPLC-UV	8	LOD (μg/L): 1.16–1.59 LOQ (μg/L): 3.52–4.81	62.0–104.3	[12]
SMZ, SIX and SDMX	FPSE	HPLC-UV	6.5	CCα (μg/kg): 114.4–116.5 CCβ (μg/kg): 104.1–118.5	93–107	[11]
9 SAs	GMeS microextraction	HPLC-DAD	30	LOQ (μg/kg): 0.31–0.91	90–105	[28]
4 SAs	PU-GO sponge microextraction	HPLC-DAD	14	LOQ: 50 (μg/kg)	90.2–112.1	This study

3. Materials and Methods

3.1. Chemicals and Reagents

Sulfathiazole (STZ), sulfamethizole (SMT), sulfadiazine (SDZ), and sulfanilamide (SN) were purchased from Sigma-Aldrich (Steinheim, Germany). HPLC grade acetonitrile and methanol obtained from Chem-Lab (Zedelgem, Belgium). Formic and acetic acid were of analytical grade and purchased from Chem-Lab (Zedelgem, Belgium) and Merck (Darmstadt, Germany) respectively. Ethanol, reagent grade (Chem-Lab, Zedelgem, Belgium) and ammonia, 25% solution (PANREAC QUIMICA SA, Barcelona, Spain) were used for the sponge optimization. Polyvinyl alcohol high molecular weight solid, (PVA 98–99 hydrolized) was purchased from A Johnson Company (New Brunswick, NJ, USA).

Graphite was purchased from Sigma Aldrich (St. Louis, MO, USA). Double-deionized water was filtered with 0.45 μm filter membrane before use.

Milk samples were collected from local market (Thessaloniki, Greece). Different fresh milk types were analyzed including skimmed (0% fat), semi-skimmed (1.5% fat), and full-fat milk (3.5% fat). All milk samples were kept refrigerated (at 4 °C) until use.

3.2. Instrumentation

Chromatographic separation and analysis were carried out on a Shimadzu HPLC system coupled to a Diode Array Detector (DAD) (Kyoto, Japan), equipped with Rheodyne 7725i 20 μL loop (Cotati, CA, USA). The system consisted of a Shimadzu LC-10 ADVP pump and a Shimadzu FCV-10ALVP solvent mixer (Kyoto, Japan). The chromatographic separation was achieved using a Merck-Lichrospher RP8e, 5 μm 250 × 4 mm analytical column (Darmstadt, Germany). Degassing of the mobile phase was performed by helium DGU-10B degassing unit by Shimadzu (Kyoto, Japan) directly in the solvent reservoirs. The system was controlled by Shimadzu LabSolutions software (Shimadzu, Kyoto, Japan) which was also used for the data acquisition and analysis.

A glass vacuum filtration apparatus obtained from Alltech Associates (Deerfield, IL, USA), was employed for the filtration of the solvents using cellulose nitrate 0.2 μm membrane filters from Whatman (Maidstone, UK) prior to use. A Glasscol Vortexer (Terre Haute, IN, USA), an ultrasonic bath Transonic 460/H (Elma, Germany), a Reacti-Vap evaporator model from PIERCE (Rockford, IL, USA), and a Hermle centrifugation (Gosheim, Germany) were acquired for the sample preparation. Moreover, a 20–200 μL micropipette ISOLAB Laborgerate GmbH (Wertheim, Germany) was used for the preparation of the standard solutions.

XRD measurements were performed on a Philips PW1820 X-ray diffractometer. The Fourier Transform Infrared Spectra (FTIR) were measured on a Nicolet 560 (Thermo Fisher Scientific Inc., MA, USA) spectrometer.

3.3. Chromatography

The mobile phase consisted of water, containing 0.1% (*v*/*v*) formic acid (A), acetonitrile (B), and methanol (C). The analytes were separated following a gradient elution program, starting at 80:3:17 (*v*/*v*/*v*), turning to 74:6:20 (*v*/*v*/*v*) in the next 7.5 min, kept isocratic for 2.5 min, and finally changing to 50:10:40 (*v*/*v*/*v*) in the last three minutes. The flow rate was set at 1.0 mL min^{-1}, while monitoring of the analytes was set at 265 nm.

3.4. Functionalization of Sponges

A commercially available polyurethane sponge was cut into cubes, immersed into ethanol/water solution, and placed in an ultrasonic bath for 20 min. The sponge was left at room temperature to dry and then it was dipped in a GO mixture for 24 h to be stirred mechanically. The mixture was prepared by the addition of 0.12 g GO in 60 mL NH_3/EtOH solution (4:1, *v*/*v*). When mechanical stirring was completed, the sponge was left to dry in room temperature. Subsequently it was rinsed with water and PVA solvent was added as a final step. The PU-GO sponge is shown in Figure 8.

Figure 8. Image of the polyurethane-graphene oxide (PU-GO) sponge.

3.5. Sample Preparation

In the present study, defatted bovine milk was used and the proteins' precipitation was achieved by adding 3 mL of ACN in 1.5 g of milk. The pH was adjusted to 5.0 by using 0.5 mL of buffer solution (70% CH_3COONa 0.2 M/30% CH_3COOH 0.2 M). The sponge was initially placed in a vial containing 1.5 g of milk and the system was centrifuged at low rpm for 15 min. The material was rinsed with deionized water and then squeezed to wash the water away. Subsequently, 1.5 mL of 1% CH_3COOH/ACN/MeOH solution (50:40:10 *v*/*v*/*v*) was added to the sponge and the analytes were eluted by centrifugation at low rpm for 15 min. The eluent was filtered and injected in the HPLC column.

In the case of fat containing milk samples, centrifugation was applied for fat removal prior to deproteinization. Moreover, sample preconcentration was applied by evaporation of elution solvent prior to HPLC analysis and reconstitution to 100 µL when necessary and in order to reach the legislation demands.

3.6. Standard Solution Preparation

For the chromatographic analysis, stock standard solutions of each analyte were prepared at a concentration of 100 ng μL^{-1} using a solvent with the same composition as the mobile phase. Stock standard solutions were stable for six months at 4 °C, while working standards were prepared on a daily basis. The calibration curves were constructed by the use of solutions being prepared at concentrations of 0.5–10 ng μL^{-1}.

3.7. Method Validation

The method was validated using spiked samples, under the optimal conditions, in terms of linearity, sensitivity, selectivity, and precision (repeatability and between-day precision), decision limit (CCa), decision capability (CCb), and stability according to the European Decision 657/2002/EC [40].

Linearity was studied by triplicate analysis of working standard solutions at concentration levels between 0.5 ng μL^{-1} to 10 ng μL^{-1}. In milk, linearity was examined by triplicate analysis of spiked samples within the range of 50 $\mu g\ kg^{-1}$–10,000 $\mu g\ kg^{-1}$ and calibration curves were calculated. Limits of detection (LOD) and quantification (LOQ) were considered as the concentration giving a signal to noise ratio of 3 and 10, respectively. The selectivity of the method was proved by the absence of interference of endogenous compounds in the analysis of blank milk samples.

Precision and accuracy were calculated by analyzing spiked samples at the concentration levels of 50 $\mu g\ kg^{-1}$, 100 $\mu g\ kg^{-1}$ and 150 $\mu g\ kg^{-1}$, which correspond to the $\frac{1}{2}$ MRL, MRL, and 1 $\frac{1}{2}$ MRL of sulfonamides [6]. Within-day repeatability was examined by 4 measurements at the above concentration levels. Between-day precision was assessed by performing triplicate analysis at the same concentration levels in three days. The relative recovery was calculated using the formula of the percentage of the ratio of the analyte mass that was found in the spiked sample, to the spiked mass.

Decision limit (CCa) was calculated using the equation CCa = MRL + 1.64 × SD, where SD is the standard deviation of the duplicate measurements of twenty milk samples spiked at MRL concentrations of each analyte. Decision capability (CCb) was calculated using the equation CCb = CCa + 1.64 × SD, with the SD being the standard deviation of the duplicate measurements of twenty milk samples spiked at CCa concentrations of each sulfonamide.

4. Conclusions

In the present study a new novel material was presented. Particularly, a PU-GO sponge was prepared, taking advantage of the unique properties of GO combined with the characteristics of the common PU sponge. This novel material was applied for the sample preparation of milk samples for the determination of sulfonamides prior to HPLC. The easy preparation of the material and the extremely fast, simple, and green sample preparation procedure make the proposed method suitable for the analysis of a complex matrix such as milk. It is the first time that the PU-GO sponge was applied for the determination of sulfonamides in milk samples. Furthermore, it is a less costly and time-consuming method and requires less equipment than previously reported methods.

Author Contributions: Conceptualization, V.F.S. and E.A.D.; methodology, software, validation, formal analysis, investigation, resources, data curation, V.F.S., E.A.D. and M.M. Writing—original draft preparation, V.F.S., E.A.D. and M.M. writing—review and editing, V.F.S. and E.A.D.; supervision, project administration, V.F.S. and E.A.D.

Funding: This research received no external funding.

Conflicts of Interest: The authors declare that there are no conflicts of interest.

References

1. Samanidou, V.F.; Tolika, E.P.; Papadoyannis, I.N. Development and validation of an HPLC confirmatory method for the residue analysis of four sulfonamides in cow's milk according to the European Union decision 2002/657/EC. *J. Liq. Chromatogr. Relat. Technol.* **2008**, *31*, 1358–1372. [CrossRef]
2. Arroyo-Manzanares, N.; Gámiz-Gracia, L.; García-Campaña, A.M. Alternative sample treatments for the determination of sulfonamides in milk by HPLC with fluorescence detection. *Food Chem.* **2014**, *143*, 459–464. [CrossRef] [PubMed]
3. Karageorgou, E.; Christoforidou, S.; Ioannidou, M.; Psomas, E.; Samouris, G. Detection of β-lactams and chloramphenicol residues in raw milk—development and application of an HPLC-DAD method in comparison with microbial inhibition assays. *Foods* **2018**, *7*, 82. [CrossRef]
4. Bitas, D.; Kabir, A.; Locatelli, M.; Samanidou, V. Food sample preparation for the determination of sulfonamides by high-performance liquid chromatography: State-of-the-art. *Separations* **2018**, *5*, 31. [CrossRef]
5. Dmitrienko, S.G.; Kochuk, E.V.; Apyari, V.V.; Tolmacheva, V.V.; Zolotov, Y.A. Recent advances in sample preparation techniques and methods of sulfonamides detection—A review. *Anal. Chim. Acta* **2014**, *850*, 6–25. [CrossRef]
6. Commission Regulation (EU) Commission Regulation (EU) No 37/2010. *Off. J. Eur. Union* **2010**, *L 15*, 1–72.
7. Kechagia, M.; Samanidou, V.; Kabir, A.; Furton, K.G. One-pot synthesis of a multi-template molecularly imprinted polymer for the extraction of six sulfonamide residues from milk before high-performance liquid chromatography with diode array detection. *Sep. Sci.* **2018**, *41*, 723–741. [CrossRef]
8. Wang, S.; Zhang, H.Y.; Wang, L.; Duan, Z.J.; Kennedy, I. Analysis of sulphonamide residues in edible animal products: A review. *Food Addit. Contam.* **2006**, *23*, 362–384. [CrossRef]
9. Zotou, A.; Vasiliadou, C. LC of sulfonamide residues in poultry muscle and eggs extracts using fluorescence pre-column derivatization and monolithic silica column. *J. Sep. Sci.* **2010**, *33*, 11–22. [CrossRef]
10. McClure, E.L.; Wong, C.S. Solid phase microextraction of macrolide, trimethoprim, and sulfonamide antibiotics in wastewaters. *J. Chromatogr. A* **2007**, *1169*, 53–62. [CrossRef]

11. Karageorgou, E.; Manousi, N.; Samanidou, V.; Kabir, A.; Furton, K.G. Fabric phase sorptive extraction for the fast isolation of sulfonamides residues from raw milk followed by high performance liquid chromatography with ultraviolet detection. *Food Chem.* **2016**, *196*, 428–436. [CrossRef]
12. Li, Y.; Wu, X.; Li, Z.; Zhong, S.; Wang, W.; Wang, A.; Chen, J. Fabrication of $CoFe_2O_4$-graphene nanocomposite and its application in the magnetic solid phase extraction of sulfonamides from milk samples. *Talanta* **2015**, *144*, 1279–1286. [CrossRef]
13. Garrido Frenich, A.; del Mar Aguilera-Luiz, M.; Martínez Vidal, J.L.; Romero-González, R. Comparison of several extraction techniques for multiclass analysis of veterinary drugs in eggs using ultra-high pressure liquid chromatography-tandem mass spectrometry. *Anal. Chim. Acta* **2010**, *661*, 150–160. [CrossRef]
14. Maggira, M.; Samanidou, V. QuEChERS: The dispersive methodology approach for complex matrices. *J. Chromatogr. Sep. Tech.* **2018**, *9*. [CrossRef]
15. Yan, H.; Sun, N.; Liu, S.; Row, K.H.; Song, Y. Miniaturized graphene-based pipette tip extraction coupled with liquid chromatography for the determination of sulfonamide residues in bovine milk. *Food Chem.* **2014**, *158*, 239–244. [CrossRef]
16. Wu, L.; Yu, L.; Ding, X.; Li, P.; Dai, X.; Chen, X.; Zhou, H.; Bai, Y.; Ding, J. Magnetic solid-phase extraction based on graphene oxide for the determination of lignans in sesame oil. *Food Chem.* **2017**, *217*, 320–325. [CrossRef]
17. Liu, Q.; Shi, J.; Zeng, L.; Wang, T.; Cai, Y.; Jiang, G. Evaluation of graphene as an advantageous adsorbent for solid-phase extraction with chlorophenols as model analytes. *J. Chromatogr. A* **2011**, *1218*, 197–204. [CrossRef]
18. Wang, X.; Liu, B.; Lu, Q.; Qu, Q. Graphene-based materials: Fabrication and application for adsorption in analytical chemistry. *J. Chromatogr. A* **2014**, *1362*, 1–15. [CrossRef]
19. Liu, Q.; Shi, J.; Sun, J.; Wang, T.; Zeng, L.; Jiang, G. Graphene and graphene oxide sheets supported on silica as versatile and high-performance adsorbents for solid-phase extraction. *Angew. Chem. Int. Ed.* **2011**, *50*, 5913–5917. [CrossRef]
20. Fumes, B.H.; Silva, M.R.; Andrade, F.N.; Nazario, C.E.D.; Lanças, F.M. Recent advances and future trends in new materials for sample preparation. *TrAC Trends Anal. Chem.* **2015**, *71*, 9–25. [CrossRef]
21. Maggira, M.; Samanidou, V.F. Graphene based materials in sample preparation prior to HPLC analysis and their applications. In *High Performance Liquid Chromatography, Types, Parameters Applications*; Ivan, L., Ed.; Nova Science Publishers, Inc.: Suite N Hauppauge, NY, USA, 2018.
22. Guan, W.; Li, Z.; Zhang, H.; Hong, H.; Rebeyev, N.; Ye, Y.; Ma, Y. Amine modified graphene as reversed-dispersive solid phase extraction materials combined with liquid chromatography-tandem mass spectrometry for pesticide multi-residue analysis in oil crops. *J. Chromatogr. A* **2013**, *1286*, 1–8. [CrossRef]
23. Huang, K.J.; Yu, S.; Li, J.; Wu, Z.W.; Wei, C.Y. Extraction of neurotransmitters from rat brain using graphene as a solid-phase sorbent, and their fluorescent detection by HPLC. *Microchim. Acta* **2012**, *176*, 327–335. [CrossRef]
24. Zhao, G.; Song, S.; Wang, C.; Wu, Q.; Wang, Z. Determination of triazine herbicides in environmental water samples by high-performance liquid chromatography using graphene-coated magnetic nanoparticles as adsorbent. *Anal. Chim. Acta* **2011**, *708*, 155–159. [CrossRef]
25. Wu, Q.; Liu, M.; Ma, X.; Wang, W.; Wang, C.; Zang, X.; Wang, Z. Extraction of phthalate esters from water and beverages using a graphene-based magnetic nanocomposite prior to their determination by HPLC. *Microchim. Acta* **2012**, *177*, 23–30. [CrossRef]
26. Zhang, X.; Niu, J.; Zhang, X.; Xiao, R.; Lu, M.; Cai, Z. Graphene oxide-SiO_2 nanocomposite as the adsorbent for extraction and preconcentration of plant hormones for HPLC analysis. *J. Chromatogr. B Anal. Technol. Biomed. Life Sci.* **2017**, *1046*, 58–64. [CrossRef]
27. Ibarra, I.S.; Miranda, J.M.; Rodriguez, J.A.; Nebot, C.; Cepeda, A. Magnetic solid phase extraction followed by high-performance liquid chromatography for the determination of sulfonamides in milk samples. *Food Chem.* **2014**, *157*, 511–517. [CrossRef]
28. Chatzimitakos, T.; Samanidou, V.; Stalikas, C.D. Graphene-functionalized melamine sponges for microextraction of sulfonamides from food and environmental samples. *J. Chromatogr. A* **2017**, *1522*, 1–8. [CrossRef]
29. Wu, Q.; Zhao, G.; Feng, C.; Wang, C.; Wang, Z. Preparation of a graphene-based magnetic nanocomposite for the extraction of carbamate pesticides from environmental water samples. *J. Chromatogr. A* **2011**, *1218*, 7936–7942. [CrossRef]

30. Su, C.; Yang, H.; Song, S.; Lu, B.; Chen, R. A magnetic superhydrophilic/oleophobic sponge for continuous oil-water separation. *Chem. Eng. J.* **2017**, *309*, 366–371. [CrossRef]
31. Su, C.; Yang, H.; Zhao, H.; Liu, Y.; Chen, R. Recyclable and biodegradable superhydrophobic and superoleophilic chitosan sponge for the effective removal of oily pollutants from water. *Chem. Eng. J.* **2017**, *330*, 423–432. [CrossRef]
32. Meng, H.; Yan, T.; Yu, J.; Jiao, F. Super-hydrophobic and super-lipophilic functionalized graphene oxide/polyurethane sponge applied for oil/water separation. *Chin. J. Chem. Eng.* **2018**, *26*, 957–963. [CrossRef]
33. Sheng, C.; Wenting, B.; Shijian, T.; Yuechuan, W. Electrochromic behaviors of poly (3-*n*-octyloxythiophene). *Polymer (Guildf).* **2008**, 1–6.
34. Kyzas, G.Z.; Travlou, N.A.; Kyzas, G.Z.; Lazaridis, N.K.; Deliyanni, E.A. Functionalization of graphite oxide with magnetic chitosan for the preparation of a nanocomposite dye. Functionalization of graphite oxide with magnetic chitosan for the preparation of a nanocomposite dye adsorbent. *Langmuir* **2015**, *29*, 1657–1668.
35. Travlou, N.A.; Kyzas, G.Z.; Lazaridis, N.K.; Deliyanni, E.A. Graphite oxide/chitosan composite for reactive dye removal. *Chem. Eng. J.* **2013**, *217*, 256–265. [CrossRef]
36. Istanbullu, H.; Ahmed, S.; Sheraz, M.A.; Rehman, I. ur Development and characterization of novel polyurethane films impregnated with tolfenamic acid for therapeutic applications. *Biomed Res. Int.* **2013**, *2013*, 1–8. [CrossRef]
37. Zhou, S.; Hao, G.; Zhou, X.; Jiang, W.; Wang, T.; Zhang, N.; Yu, L. One-pot synthesis of robust superhydrophobic, functionalized graphene/polyurethane sponge for effective continuous oil-water separation. *Chem. Eng. J.* **2016**, *302*, 155–162. [CrossRef]
38. Xu, Y.; Hong, W.; Bai, H.; Li, C.; Shi, G. Strong and ductile poly(vinyl alcohol)/graphene oxide composite films with a layered structure. *Carbon N. Y.* **2009**, *47*, 3538–3543. [CrossRef]
39. Liang, J.; Huang, Y.; Zhang, L.; Wang, Y.; Ma, Y.; Cuo, T.; Chen, Y. Molecular-level dispersion of graphene into poly(vinyl alcohol) and effective reinforcement of their nanocomposites. *Adv. Funct. Mater.* **2009**, *19*, 2297–2302. [CrossRef]
40. Commission Decision, 2002/657/EC. 2002, pp. 8–36. Available online: https://eur-lex.europa.eu/legal-content/EN/ALL/?uri=CELEX%3A32002D0657 (accessed on 31 May 2019).
41. Tolmacheva, V.V.; Apyari, V.V.; Furletov, A.A.; Dmitrienko, S.G.; Zolotov, Y.A. Facile synthesis of magnetic hypercrosslinked polystyrene and its application in the magnetic solid-phase extraction of sulfonamides from water and milk samples before their HPLC determination. *Talanta* **2016**, *152*, 203–210. [CrossRef]
42. Hou, X.; Chen, G.; Zhu, L.; Yang, T.; Zhao, J.; Wang, L.; Wu, Y. Development and validation of an ultra high performance liquid chromatography tandem mass spectrometry method for simultaneous determination of sulfonamides, quinolones and benzimidazoles in bovine milk. *J. Chromatogr. B* **2014**, *962*, 20–29. [CrossRef]

Sample Availability: Samples of the compounds are not available from the authors.

Article

Assessing Geographical Origin of *Gentiana Rigescens* Using Untargeted Chromatographic Fingerprint, Data Fusion and Chemometrics

Tao Shen [1,2,3], Hong Yu [1,2,*] and Yuan-Zhong Wang [4]

1 Yunnan Herbal Laboratory, Institute of Herb Biotic Resources, School of Life and Sciences, Yunnan University, Kunming 650091, China
2 The International Joint Research Center for Sustainable Utilization of Cordyceps Bioresouces in China and Southeast Asia, Yunnan University, Kunming 650091, China
3 College of Chemistry, Biological and Environment, Yuxi Normal University, Yu'xi 653100, China
4 College of Traditional Chinese Medicine, Yunnan University of Chinese Medicine, Kunming 650500, China
* Correspondence: hongyu@ynu.edu.cn or herbfish@163.com; Tel.: +86-0871-68182671

Academic Editor: Marcello Locatelli
Received: 10 June 2019; Accepted: 12 July 2019; Published: 14 July 2019

Abstract: *Gentiana rigescens* Franchet, which is famous for its bitter properties, is a traditional drug of chronic hepatitis and important raw materials for the pharmaceutical industry in China. In the study, high-performance liquid chromatography (HPLC), coupled with diode array detector (DAD) and chemometrics, were used to investigate the chemical geographical variation of *G. rigescens* and to classify medicinal materials, according to their grown latitudes. The chromatographic fingerprints of 280 individuals and 840 samples from rhizomes, stems, and leaves of four different latitude areas were recorded and analyzed for tracing the geographical origin of medicinal materials. At first, HPLC fingerprints of underground and aerial parts were generated while using reversed-phase liquid chromatography. After the preliminary data exploration, two supervised pattern recognition techniques, random forest (RF) and orthogonal partial least-squares discriminant analysis (OPLS-DA), were applied to the three HPLC fingerprint data sets of rhizomes, stems, and leaves, respectively. Furthermore, fingerprint data sets of aerial and underground parts were separately processed and joined while using two data fusion strategies ("low-level" and "mid-level"). The results showed that classification models that are based OPLS-DA were more efficient than RF models. The classification models using low-level data fusion method built showed considerably good recognition and prediction abilities (the accuracy is higher than 99% and sensibility, specificity, Matthews correlation coefficient, and efficiency range from 0.95 to 1.00). Low-level data fusion strategy combined with OPLS-DA could provide the best discrimination result. In summary, this study explored the latitude variation of phytochemical of *G. rigescens* and developed a reliable and accurate identification method for *G. rigescens* that were grown at different latitudes based on untargeted HPLC fingerprint, data fusion, and chemometrics. The study results are meaningful for authentication and the quality control of Chinese medicinal materials.

Keywords: authentication; liquid chromatography fingerprint; chemometrics; random forest; OPLS-DA; data fusion; *Gentiana rigescens*

---

1. Introduction

Gentiana rigescens Franchet (Dian long dan) is a herbaceous species that grows in mountainous regions of Yunnan-Guizhou Plateau in the southwest of China [1]. Like European traditional medicinal plant yellow gentian (*G. lutea* L), *G. rigescens* is famous for its bitter properties that are due to the bitter active principles (e.g., loganin, gentiopicroside, swertiamarin, sweroside, etc.) [2–4]. Those compounds have pharmacological effects of anti-inflammation, antioxidant, anti-cancer, antiviral, cholagogic agent, hepatoprotective, wound-healing activities, and so forth [3,5]. Additionally, they are used to stimulate appetite and improve digestion [5–7]. In addition, a series of neuritogenic compounds had been isolated from the aerial and underground parts of *G. rigescens*, which could be used as raw material for the preparation of functional food and a therapeutic drug for Alzheimer's disease [8–11]. Now, *G. rigescens* have been the official drug of Chinese pharmacopoeia (2015 edition) for chronic hepatitis and important raw materials for the pharmaceutical industry in China [12].

G. rigescens were usually collected from different regions of Yunnan-Guizhou Plateau in order to provide satisfaction of continuously increasing industrial demands for raw materials. However, some of the researchers had reported that chemical constitutions of underground part of *G. rigescens* were extremely variable and diverse according to plant grown location or producing area [13–15]. Quantitative analysis of bioactivity compounds (such as gentiopicroside, sweroside, swertiamarin, isoorientin, and other compounds) from rhizomes, stems, leaves, and flowers indicated that northwest of Yunnan-Guizhou Plateau was suitable for chemical compounds accumulation [13–16]. Additionally, conversion and transport of those compounds might be influenced by climatic conditions in the plant habitat [14,17].

Latitude has a strong impact on the local climate environment in southwest China [18,19]. As the main distribution area of *G. rigescens*, Yunnan-Guizhou Plateau is characterized by very complex topography and it displays a wide variety of micro-climates [18–21]. There are six climatic zones from the north towards the south [20]. Especially, in the higher latitude areas, such as northwest Yunnan or south of the Hengduan Mountains (26–28° N), the temperature gradients are more abrupt than in the other regions [19]. Furthermore, precipitation and temperature in the Yunnan-Guizhou Plateau also show clear variations along the latitude gradients [19,21]. Therefore, it is necessary to explore the variation of phytochemical and medicinal material quality of *G. rigescens* that were grown in different latitudes and build a classification model for tracing producing areas of medicinal materials.

As we know, the contents of bioactive compounds and quality of medicinal materials have a close relationship with the environment of producing area [22–25]. Quality control and geographical indication of medicinal materials raise many concerns by pharmaceutical industries with the expansion in the use of herbal medicines. However, using few marker compounds could not reflect the chemical complexity of herbs and this method is hard to effectively authenticate the origin of herbal medicines [26,27]. Chemical fingerprints, as a comprehensive evaluation methodology, have been widely used to deal with the problem [26,28,29]. In recent years, infrared spectroscopy (IR), UV-Vis spectroscopy (UV-Vis), and other spectral fingerprints have been well-established analytical techniques for geographical traceability studies of *G. rigescens* and other medicinal plants in the worldwide [30–34]. In contrast, there were limited reports on the use of chromatographic fingerprint to identify the producing regions of herbal materials [30–35]. Although there were many reports about discrimination of herbs according to their producing areas while using liquid chromatography technology, most of them are based on the information of limited chemical markers or chromatographic profiles [36–39]. The potential of chromatographic fingerprints for herbs authentication needs to be further explored.

When compared with chemical marker or chromatographic profile (targeted), chromatographic fingerprint (untargeted) contains unspecific and non-evident information and chemometric tools should extract chemical information [40]. Recently, literature reported some successful studies applying chromatographic fingerprint, together with chemometric methodology, to discriminate herbs and food samples of different origin or cultivars [41–44]. All of those studies suggested that it is possible to develop a reliable and accurate method for the geographical tracing of *G. rigescens* by applying the chromatographic fingerprint methodology.

In the progression of improving geographical authentication of food and drugs, one of the important goals is building discrimination models with a less error rate and reducing the uncertainty of the prediction results [33,44]. Data fusion strategy has been widely used in the last years in the field of food authentication in order to improve class discrimination techniques [45]. Some reports about *Panax notoginseng*, *Paris Polyphylla* var. *yunnanensis* and other herb materials also showed the huge potential of this strategy in the discrimination of medicinal materials producing areas [46–48]. Today, most of the fused data come from spectral fingerprint and very few studies report the data fusion of chromatographic fingerprint [42,43]. Furthermore, data fusion studies are mostly based on the fusion of multivariate instrumental techniques [42,43], while reports of *P. Polyphylla* var. *yunnanensis*, *Macrohyporia cocos*, and other species indicated that reliable classification results were also available by the fusion analysis of chemical fingerprint data collected from different medicinal parts of herbs [35,49]. Accumulation and distribution of metabolites in the different parts of plants were different because of the differential response of root, stem, flower and other organs to the environment variation of producing area [17,50]. Therefore, fingerprint data fusion of multi-medicinal parts may provide integrated chemical information for the authentication of medicinal materials. At the same time, this method also contributes to a more comprehensive understanding of the response and adaptation of medicinal plants to complex geographical environments.

The aim of this study is to explore the variation of chromatographic fingerprints of *G. rigescens* along the latitude gradients and to use chemometrics to mine fingerprint chemical information, and to investigate the potential of the untargeted chromatographic fingerprint to trace herbs grown at different latitudes. For this purpose, we developed fingerprint of rhizomes, stems, and leaves of *G. rigescens* by high-performance liquid chromatography with diode array detection (HPLC-DAD) technology. Subsequently, classification models for the identification of different producing areas were built by HPLC fingerprint combined with RF (random forest algorithm) and OPLS-DA (orthogonal partial least-squares discriminant analysis). At last, two types of data fusion strategies, "low- level" and "mid-level" data fusion, were studied in order to improve the model performances.

2. Results and Discussion

2.1. Chromatographic Fingerprints Variation Along the Latitude Gradients

Figure 1 displays the representative chromatographic fingerprints of rhizome, stem, and leaf. From HPLC fingerprints, it can be found that the five marker compounds of iridoids were eluted before 15 min. The retention times (*t*/min) of loganin (1), 6′-*O*-β-D-glucopyranosylgentiopicroside (2), swertiamarine (3), gentiopicroside (4), and sweroside (5) were 7.279, 9.213, 9.573, 11.376, and 11.622 min, respectively. Loganin and gentiopicroside were mainly accumulation in the underground part and sweroside accumulated more in the overground parts. Furthermore, differences in the chemical composition of rhizome, stem, and leaf can also be visually observed through chromatographic fingerprints. For facilitating subsequent data exploration and modeling analysis, the retention time of fingerprints signal was replaced by variables (Figure 1d–f). As a result, there were 3839, 4140, and 4140 variables of rhizome, stem, and leaf fingerprints, respectively.

Principal component analysis (PCA) and two-dimensional score plots visualized the differences and variation trends of three medicinal parts. Figure 2 shows that the rhizomes and stems of *G. rigescens* tended to cluster to the left part, while the leaves data scattered to the right.

Although the fingerprints between the aboveground and underground medicinal parts were obvious differences, an interesting result is that a trend of separation according to product region latitude was observed from the PCA and score plots of samples of three medicinal parts. For example, two-dimensional score plots of chromatographic fingerprint of rhizomes showed that the samples separation trend increases with an increase in geographical distance and a clear separation between samples that were collected from lower latitude and higher latitude regions (Figure 3). In contrast to this, when considering the separation between samples with product regions geographically close to each other, we observed that the rhizome samples separation trend decreases with a decrease in the geographical distance (Figure 4). The PCA score plots of stems and leaves changed in the same trend as rhizomes (Figures S1–S4).

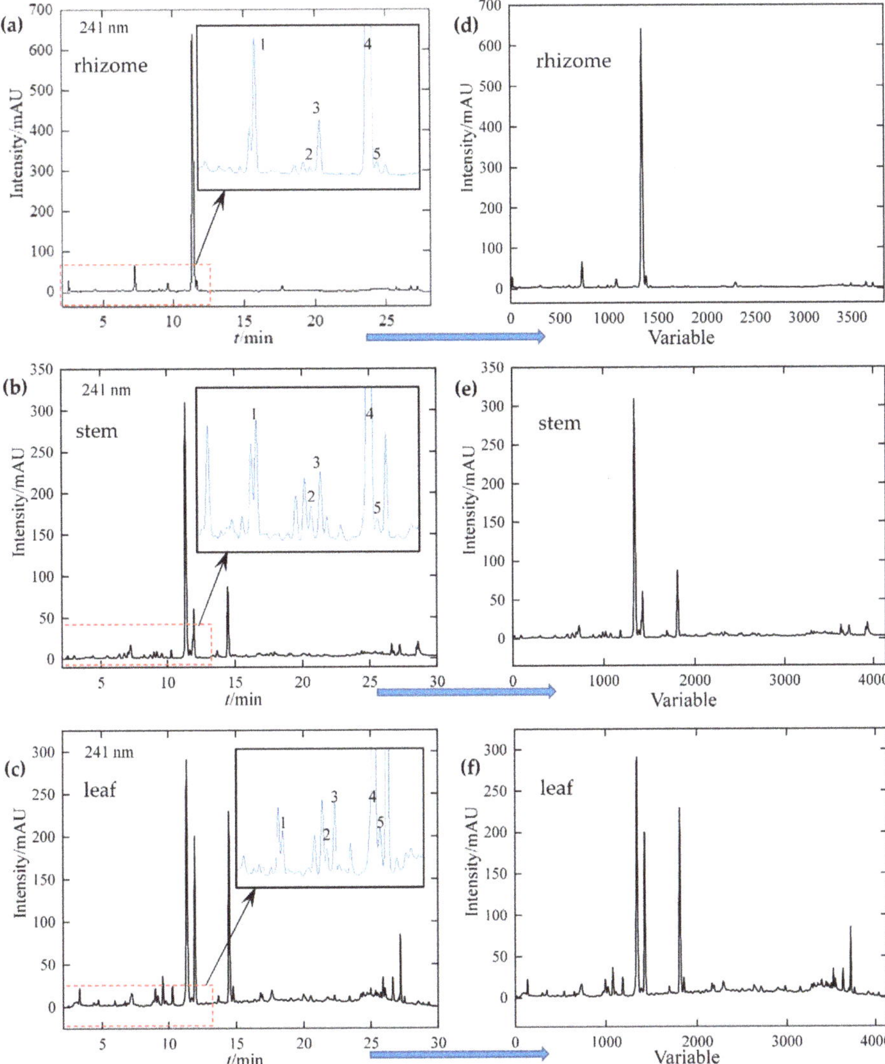

Figure 1. High-performance liquid chromatography (HPLC) fingerprint of rhizome (**a**), stem (**b**), leaf (**c**) and fingerprints after variable transformation (**d**–**f**). (1) loganin, (2) 6′-*O*-β-D-glucopyranosylgentiopicroside, (3) swertiamarine, (4) gentiopicroside, and (5) sweroside.

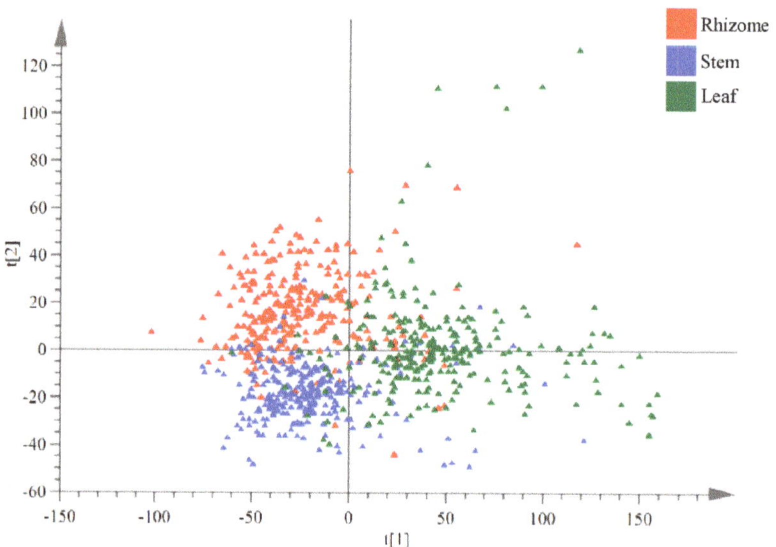

Figure 2. Two-dimensional principal component score plot of rhizomes, stems, and leaves samples based on chromatographic fingerprint data.

The results of PCA highlighted that the chromatographic fingerprints of *G. rigescens* were different among rhizomes, stems, and leaves, and were affected by latitude gradients of the production regions. Especially between lower latitudes and higher latitudes, the samples seem to be clearly distinguishable. Based on PCA exploratory analysis (unsupervised methods), supervised pattern recognition (OPLS-DA) should be applied to gain better classification results for samples that were grown in different latitudes (Figures 5 and 6), and OPLS-DA and variable importance in the projection (VIP) analysis were used to further investigate the fingerprint variables of *G. rigescens* that were sensitive to latitude changes.

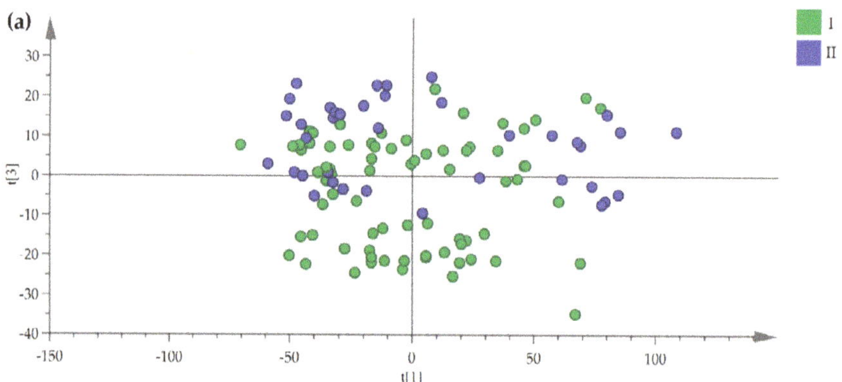

Figure 3. *Cont.*

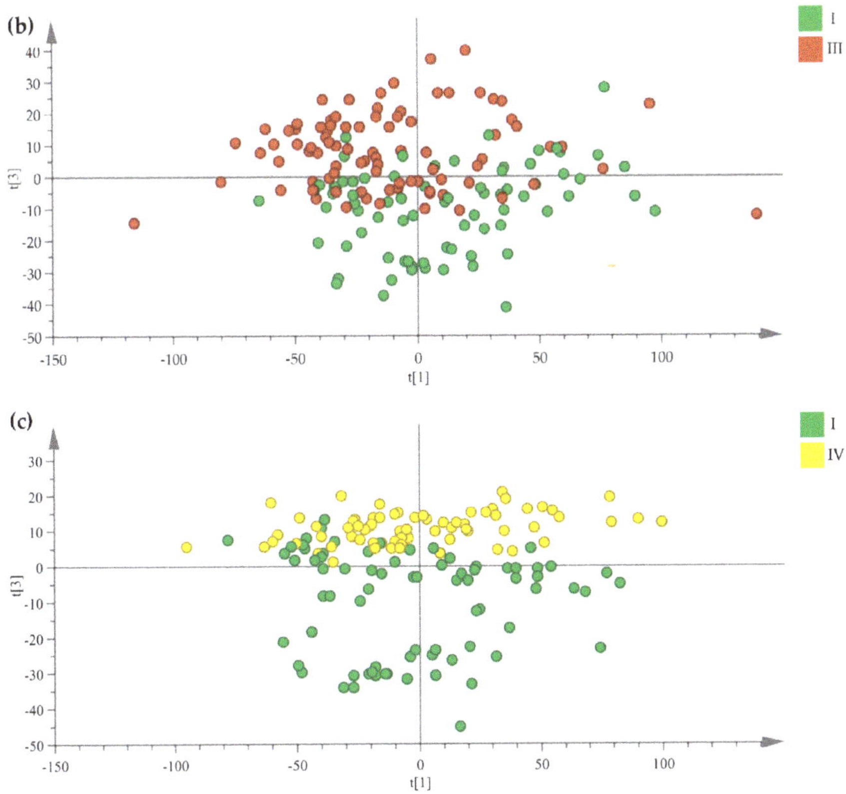

Figure 3. Variation of rhizomes score plots along the latitude gradients. (**a**) is low latitude and mid-latitude, (**b**) is low latitude and mid-high latitude and (**c**) is low latitude and high latitude (green circles = low latitudes area, 23.92–23.66° N, blue circles = mid-latitude area, 24.95–25.06° N, red circles = mid-high latitude area, 26.49–26.64° N, yellow circles = high latitude area, 27.34–28.52° N).

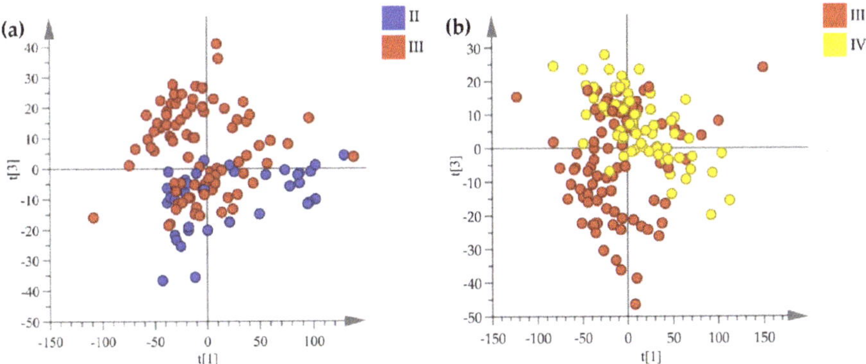

Figure 4. Variation of rhizomes score plots between the adjacent latitudes. (**a**) is mid-latitude and mid-high latitude and (**b**) is mid-high latitude and high- latitude (blue circles = mid-latitude area, 24.95–25.06° N, red circles = mid-high latitude area, 26.49–26.64° N, yellow circles = high latitude area, 27.34–28.52° N).

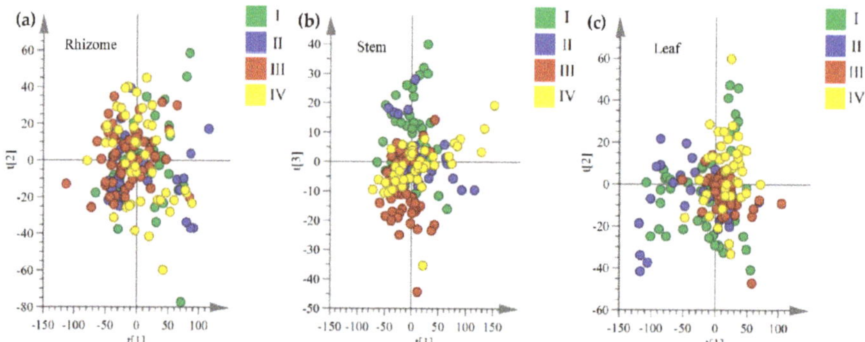

Figure 5. Two-dimensional principal component score plots for samples of rhizomes (**a**), stems (**b**), and leaves (**c**) of *G. rigescens* grown at four latitudes.

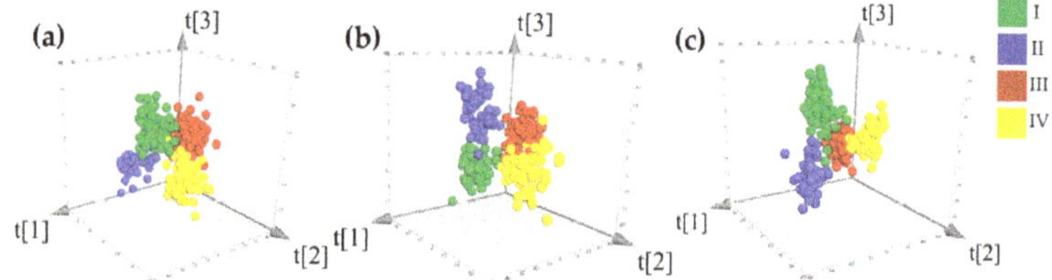

Figure 6. Three-dimensional (3D) Scores-plot diagram of rhizomes (**a**), stems (**b**), and leaves (**c**) orthogonal partial least-squares discriminant analysis (OPLS-DA) analysis among four different latitudes (OPLS-DA model (**a**) $R^2 = 0.74$ and $Q^2 = 0.68$, model (**b**) $R^2 = 0.75$ and $Q^2 = 0.68$, model (**c**) $R^2 = 0.72$ and $Q^2 = 0.71$, permutation plot of three models were shown in Figures S5–S7).

The variable's VIP value was greater than 1.00, which indicates that the variable was obviously affected by the change of the latitude of the producing areas. From Figure 7a, it could be found that the change of three ranges of rhizome's fingerprint was closely related to producing areas latitude. The first range was related to variables of retention time at 2.00–13.00 min. The second range was related to variables of retention time at 15.00–20.00 min. Additionally, the third range was related to the variables of retention time after 25.00 min. Figure 7b showed that important variables (VIP value > 1.00) of stem fingerprint relate to the variables of retention time at 2.00–20.00 min and 25.00–30.00 min. For leaf fingerprint, chromatographic variables, retention time at 2.00–15.00 min, 17.00–19.00 min and 25.00–30.00 min, were the most sensitive to latitude changes of producing areas (Figure 7c). According to the identification of the major compounds in fingerprint, it showed that many of these important variables were chromatographic signals of iridoids and secoiridoids, such as loganin, 6′-*O*-β-D-glucopyranosylgentiopicroside, swertiamarine, gentiopicroside, and sweroside. A previous study regarding the spatial profiling of iridoids phytochemical constituents found that the geographical variation of those compounds could be attributed to some environmental factors [13,17], for example, the difference of precipitation of natural habitats [17]. Additionally, it was interesting to note that the number of important variables after 25 min is gradually increasing from the rhizome to the leaves. The results suggested that, in addition to iridoids, other low polarity products in *G. rigescens* have implications for the differentiation of different geographical origins.

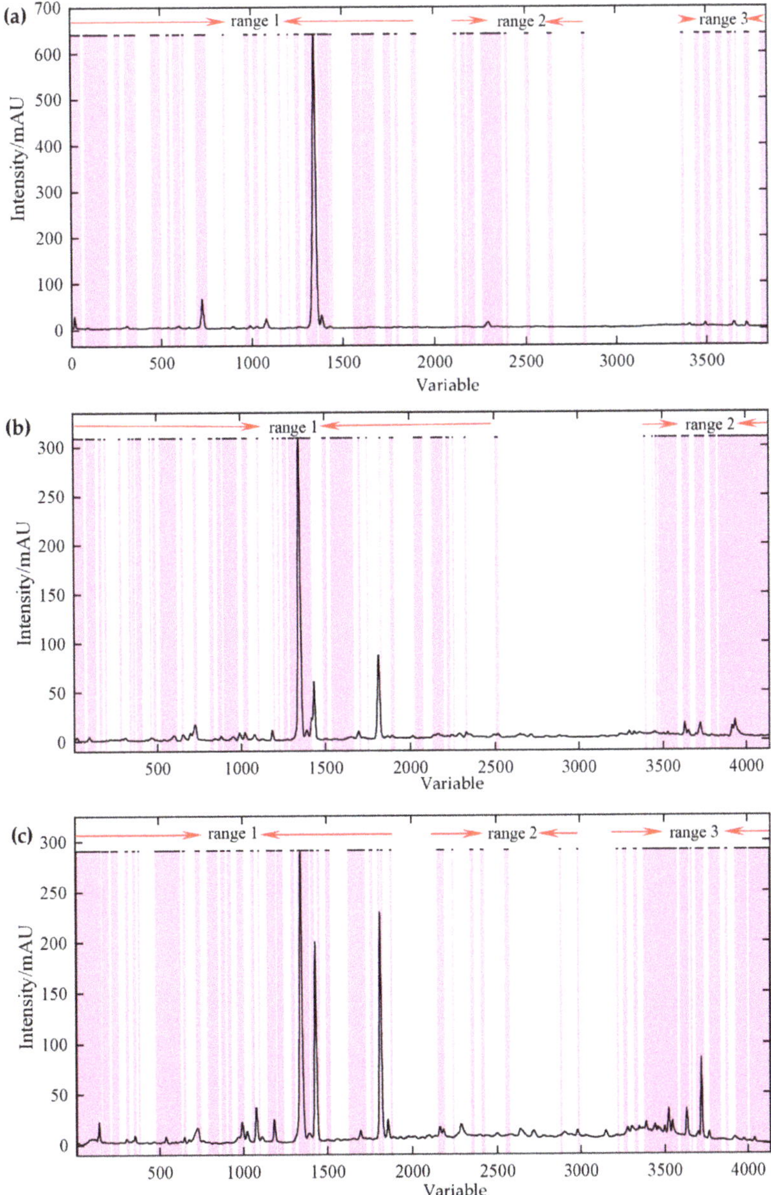

Figure 7. Important variables of fingerprint (purple = variable VIP value > 1) (**a**) rhizome, (**b**) stem, and (**c**) leaf.

In a word, current research indicated that the chemical composition of *G. rigescens* changes with the grown latitude in a way that could be traced with the chromatographic fingerprint. Furthermore, three-dimensional (3D) score plots and VIP analysis showed a difference of phytochemical geographic variation for overground and underground parts. Those differences might affect the result of geographical origin traceability of samples.

2.2. Geographic Authentication Based on Fingerprints of Different Medicinal Parts

In recent years, literature had already reported satisfying classification results that were obtained by RF or OPLS-DA models [51–54]. As an ensemble learning method, the RF algorithm could correct for decision trees' habit of overfitting to their training set [55]. Additionally, OPLS could help to overcome these obstacles by separating useful information from noise and improve complex chemical data features and interpretability [56,57]. In this work, we tested RF and OPLS-DA models, combined with rhizome, stem, and leaf fingerprint data in order to classify *G. rigescens* according to their grown latitude.

2.2.1. RF Classification

In the beginning, samples from the data set of rhizomes (280 samples and 3839 variables) were separated into a calibration set (186 samples) and a validation set (94 samples) by the Kennard-Stone algorithm. Subsequently, 186 rhizome samples that were collected from four latitude gradients were used to establish the calibration model (R_RF). During the modeling process, the initial value of n_{tree} (needs to be optimized) was defined as 2000, the initial value of m_{try} was defined as the square root of the number of variables, and the rest of the parameters were defined as the default value. Subsequently, OOB errors were calculated and the value of the best n_{tree} was obtained according to the lowest OOB error. Figure 8 shows that the minimum error and the standard error are the lowest, with 663 trees. Based on the optimal number of trees, m_{try} was re-selected by searching the values ranged from 50 to 75. The calculation results found that the m_{try} value should be defined as 61, because of the model had the lowest OOB classification error. Finally, a final classification model was established based on optimum n_{tree} and m_{try} values.

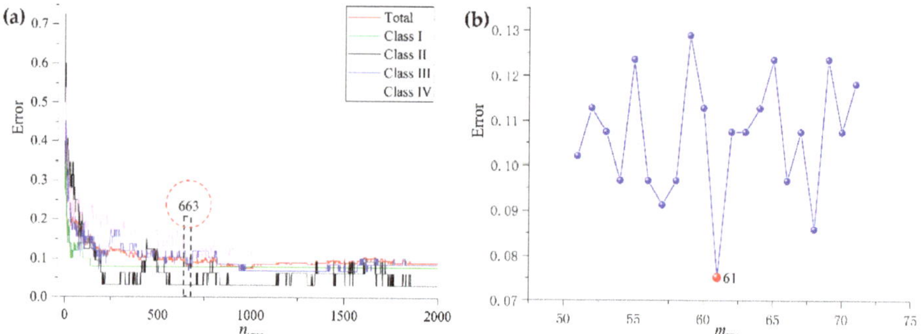

Figure 8. The n_{tree} (**a**) and m_{try} (**b**) screening of RF models based on rhizomes fingerprints.

Table 1 shows that the accuracies for samples of calibration set were 96.77% for low latitude samples, 99.46% for mid-latitude samples, 94.62% for mid-high latitude samples, and 94.09% for high latitude samples. Additionally, the accuracies of samples of validation set were 91.49%, 95.74%, 94.68%, and 98.94% for four different latitudes samples, respectively.

Table 1. The major parameters of random forest (RF) model based on rhizomes data set.

Model	Performance	Calibration Set				Validation Set			
		I	II	III	IV	I	II	III	IV
R_RF	ACC (%)	96.77	99.46	94.62	94.09	91.49	95.74	94.68	98.94
	SE	0.92	0.97	0.93	0.89	0.92	0.75	0.93	0.96
	SP	0.99	1.00	0.95	0.96	0.91	1.00	0.95	1.00
	MCC	0.92	0.98	0.88	0.84	0.80	0.84	0.88	0.97
	EFF	0.95	0.98	0.94	0.92	0.92	0.87	0.94	0.98

Like previous investigations of the rhizome model, the data set of stems (280 samples and 4140 variables) and leaves (280 samples and 4140 variables) were separated into calibration sets and validation sets, respectively. Subsequently, RF calibration modes of stems (S_RF) and leaves (L_RF) were built. The optimum n_{tree} and m_{try} could be found in Figures 9 and 10.

For the RF model of the stem, the accuracies of samples of calibration set of 92.47%, 94.62%, 93.01%, and 93.01% were achieved for low latitudes, mid-latitudes, mid-high latitudes, and high latitudes. Additionally, the accuracies of samples of validation set were 98.94%, 97.87%, 96.81%, and 97.87%, respectively (Table 2).

For RF model of the leaf, accuracies of 92.47%, 96.24%, 93.01%, and 94.62% were achieved for the calibration set. Additionally, accuracies of 85.11%, 93.62%, 89.36%, and 93.62% for the validation set (Table 3).

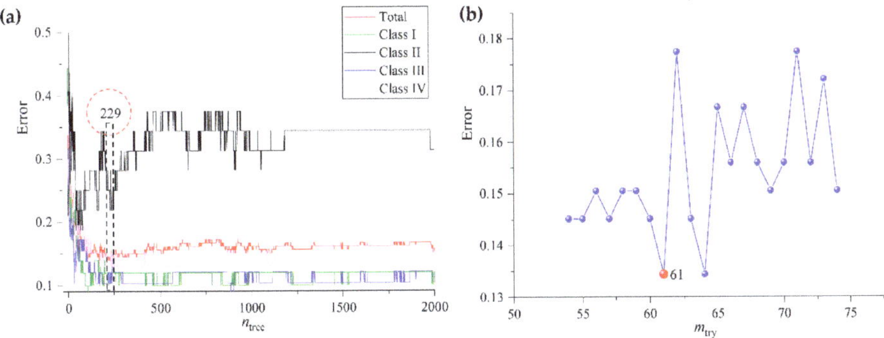

Figure 9. The n_{tree} (**a**) and m_{try} (**b**) screening of RF models based on stems fingerprints.

Table 2. The major parameters of RF model based on stems data set.

Model	Performance	Calibration Set				Validation Set			
		I	II	III	IV	I	II	III	IV
S_RF	ACC (%)	92.47	94.62	93.01	93.01	98.94	97.87	96.81	97.87
	SE	0.92	0.69	0.91	0.87	1.00	0.88	1.00	0.91
	SP	0.93	1.00	0.94	0.95	0.99	1.00	0.95	1.00
	MCC	0.82	0.80	0.84	0.81	0.97	0.92	0.93	0.94
	EFF	0.92	0.83	0.93	0.91	0.99	0.94	0.98	0.96

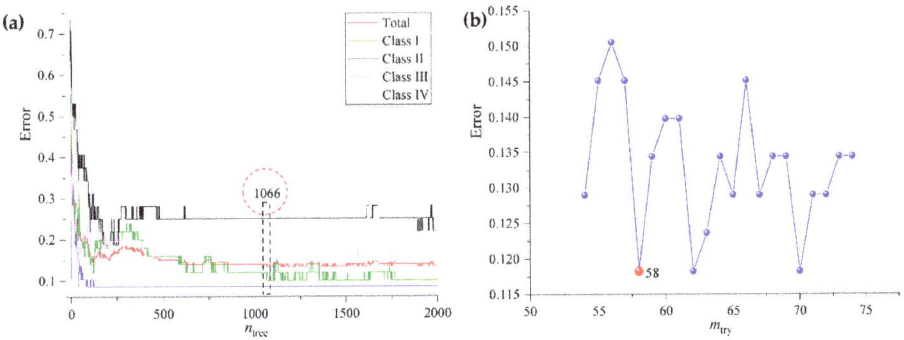

Figure 10. The n_{tree} (**a**) and m_{try} (**b**) screening of RF models based on leaves fingerprints.

Table 3. The major parameters of RF model based on leaves data set.

Model	Performance	Calibration Set				Validation Set			
		I	II	III	IV	I	II	III	IV
L_RF	ACC (%)	92.47	96.24	93.01	94.62	85.11	93.62	89.36	93.62
	SE	0.94	0.78	0.91	0.85	0.88	0.69	0.86	0.74
	SP	0.92	1.00	0.94	0.98	0.84	0.99	0.91	1.00
	MCC	0.82	0.86	0.84	0.85	0.67	0.76	0.76	0.83
	EFF	0.93	0.88	0.93	0.91	0.86	0.82	0.88	0.86

2.2.2. OPLS-DA Classification

The OPLS-DA models of rhizomes (R_OPLS-DA), stems (S_OPLS-DA), and leaves (L_OPLS-DA) were constructed based on the same calibration and validation sets that were used in RF models. All of the models were constructed based on the internal seven-fold cross-validation and permutation plot could be found in Supplementary Materials.

Table S1 showed that the R^2 of models ranged from 0.77 to 0.82 and the Q^2 of models were larger than 0.50, which indicated that the OPLS-DA models were well fitted and better predictive. The permutation test results could be found in Figures S14–S16.

The classification results of R_OPLS-DA model showed (Table 4) accuracies of calibration set were 98.92% for all classes. Accuracies of validation set were 95.47%, 98.94%, 94.86%, and 97.87% for low latitudes, mid-latitudes, mid-high latitudes, and high latitudes samples, respectively. For S_OPLS-DA models (Table 4), although 98.92%, 99.46%, 98.92%, and 98.39% values of calibration set accuracies were obtained for samples that were grown in four different latitudes, a lower value of total accuracy rate of validation set was obtained (93.62%). Parameters of L_OPLS-DA model showed (Table 4) that the accuracies of the calibration set were 97.31%, 99.46%, 97.31%, and 98.39% for low latitude, mid-latitude, mid-high latitude, and high latitude samples, respectively. However, the total accuracy of the validation set was lower than the calibration set. Especially, for samples of class 1, the accuracy was only 88.30%.

Table 4. The major parameters of OPLS-DA models.

Model	Performance	Calibration Set				Validation Set			
		I	II	III	IV	I	II	III	IV
R_OPLS-DA	ACC (%)	98.92	98.92	98.92	98.92	95.74	98.94	94.68	97.87
	SE	0.98	0.97	0.98	0.98	0.92	0.94	0.93	0.96
	SP	0.99	0.99	0.99	0.99	0.97	1.00	0.95	0.99
	MCC	0.97	0.96	0.97	0.97	0.89	0.96	0.88	0.94
	EFF	0.99	0.98	0.99	0.99	0.95	0.97	0.94	0.97
S_OPLS-DA	ACC (%)	98.92	99.46	98.92	98.39	91.49	93.62	91.49	97.87
	SE	1.00	0.97	1.00	0.93	0.92	0.81	0.83	0.91
	SP	0.99	1.00	0.98	1.00	0.91	0.96	0.95	1.00
	MCC	0.97	0.98	0.98	0.96	0.80	0.77	0.80	0.94
	EFF	0.99	0.98	0.99	0.97	0.92	0.88	0.89	0.96
L_OPLS-DA	ACC (%)	97.31	99.46	97.31	98.39	88.30	95.74	92.55	93.62
	SE	0.94	1.00	0.93	1.00	0.81	0.88	0.90	0.83
	SP	0.99	0.99	0.99	0.98	0.91	0.97	0.94	0.97
	MCC	0.93	0.98	0.94	0.96	0.71	0.85	0.83	0.82
	EFF	0.96	1.00	0.96	0.99	0.86	0.92	0.92	0.90

Finally, we made a comprehensive comparison to the six models' classification performance superiority on the basis of the above analysis. For the RF model, the order of calibration total accuracy was as follows: R_RF (96.24%) > L_RF (94.09%) > S_RF (93.28%). The order of validation total accuracy was as follows: S_RF (97.87%) > R_RF (95.21%) > L_RF (90.43%). For the OPL-DA model, the order of calibration total accuracy was as follows: R_OPL-DA (98.92%) and S_OPLS-DA (98.92%) > L_OPLS-DA (98.12%). The order of validation total accuracy was as follows: R_OPL-DA (96.81%) > S_OPLS-DA (93.62%) > L_OPLS-DA (92.55%). Classification models that were built by using leaf data set presented the worst performance from the accuracy point of view. Additionally, validation sets of the L_RF and L_OPL-DA model had lower Matthews correlation coefficient (MCC) values. By contrast, all of the models based on rhizome data set presented a better classification performance (total accuracy ranged from 95.21% to 98.92%). The best total accuracy occurred when rhizome data combined with the OPLS algorithm. We could find that phenomenon of imbalance category recognition in R_OPLS-DA model was better than other models from SE values, SP values, MCC values, and EFF value.

Although the classification performance for OPLS-DA and RF models on the basis of rhizome data set was good, the model classification ability, accuracy, sensitivity (SE), specificity (SP), MCC, and efficiency (EFF), need to be enhanced. In a further step, the feasibility of combining the information from rhizome, stem, and leaf fingerprint data for samples geographical traceability was investigated by low-level and mid-level data fusion strategies.

2.3. Geographic Authentication Based on Data Fusion Strategy

2.3.1. Low-Level Data Fusion

According to the method that was described in data preprocessing (Figure 11), fingerprint data sets of overground and underground organs as subsets were used to concatenate into a single data block (a new data set). In the case of the low-level strategy, four data sets, rhizome combined with stem (RS), rhizome combined with leaf (RL), stem combined with leaf (SL), and all data combined (RSL), were used to build RF (RS_RF, RL_RF, SL_RF, and RSL_RF) and OPLS-DA (RS_OPLS-DA, RL_OPLS-DA, SL_OPLS-DA, and RSL_OPLS-DA) models. For every data set, the samples were randomly selected as a calibration set and the rest of the samples were used as a validation set (finished by Kennard-Stone algorithm).

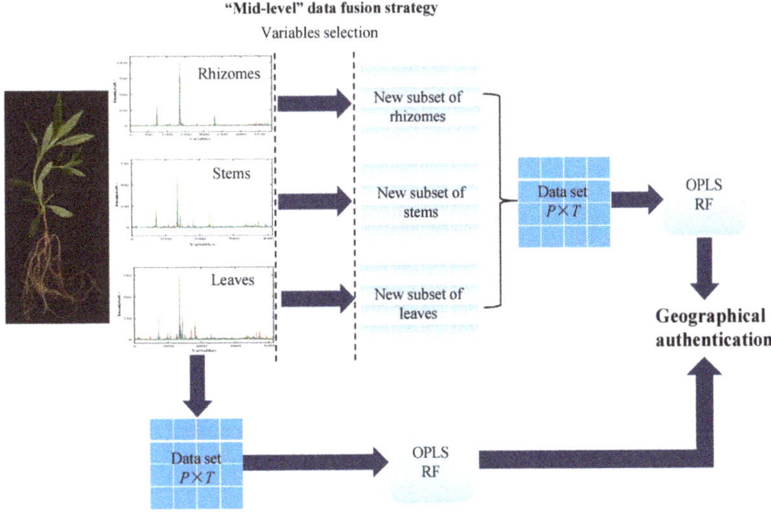

Figure 11. The workflow of geographical authentication of *G. rigescens* grown at different latitudes using data fusion strategy.

The optimum n_{tree} and m_{try} values were selected at first (Figure S8). Afterwards, final classification models were established based on the best values of arguments. From Table 5, it could be seen that the samples collected from four different latitudes were better discriminated by using RS data set and RSL data set. RS_RF model achieved 95.43% total accuracy for the calibration set and achieved 96.81% total accuracy for calibration set. RSL_RF model achieved 94.89% correctly for the calibration set and achieved 97.37% correctly for the calibration set. From a comparison with SE, SP, MCC, and EFF values of S_RF and L_RF models (Tables 1 and 3), we found that the low-level data fusion strategy improved the phenomenon of imbalance category recognition in the RF model (Table 5). However, the total accuracy of models was not obviously improved.

Table 5. The major parameters of RF models based on low-level data fusion strategy.

Model	Class	Calibration Set				Validation Set			
		I	II	III	IV	I	II	III	IV
RS_RF	ACC (%)	96.77	98.92	92.47	93.55	95.74	97.87	95.74	97.87
	SE	0.92	0.94	0.91	0.87	0.92	0.88	0.97	0.96
	SP	0.99	1.00	0.93	0.96	0.97	1.00	0.95	0.99
	MCC	0.92	0.96	0.83	0.83	0.89	0.92	0.90	0.94
	EFF	0.95	0.97	0.92	0.91	0.95	0.94	0.96	0.97
RL_RF	ACC (%)	94.09	98.39	93.01	96.24	87.23	94.68	91.49	92.55
	SE	0.90	0.91	0.91	0.91	0.96	0.75	0.86	0.70
	SP	0.96	1.00	0.94	0.98	0.84	0.99	0.94	1.00
	MCC	0.85	0.94	0.84	0.90	0.74	0.80	0.80	0.80
	EFF	0.93	0.95	0.93	0.95	0.90	0.86	0.90	0.83
SL_RF	ACC (%)	93.55	95.70	92.47	93.55	90.43	96.81	96.81	94.68
	SE	0.94	0.75	0.91	0.85	0.92	0.88	0.93	0.83
	SP	0.93	1.00	0.93	0.96	0.90	0.99	0.98	0.99
	MCC	0.84	0.84	0.83	0.82	0.78	0.88	0.92	0.85
	EFF	0.94	0.87	0.92	0.90	0.91	0.93	0.96	0.90
RSL_RF	ACC (%)	95.70	99.46	91.94	92.47	94.68	96.81	100.00	97.87
	SE	0.94	0.97	0.86	0.85	0.86	1.00	1.00	0.96
	SP	0.96	1.00	0.95	0.95	0.98	0.96	1.00	0.99
	MCC	0.89	0.98	0.81	0.80	0.87	0.88	1.00	0.94
	EFF	0.95	0.98	0.90	0.90	0.92	0.98	1.00	0.97

The permutation plot of all models could be found in Supplementary Materials (Figures S17–S20). The classification results of OPLS-DA models based on low-level data fusion showed models' R^2 values ranged from 0.86 to 0.90 and Q^2 values ranged from 0.74 to 0.80 (Table S2). Total accuracy rates of the calibration set of RS_OPLS-DA, RL_OPLS-DA, SL_OPLS-DA, and RSL_OPLS-DA were 99.46%, 99.73, 100.00%, and 99.73%, respectively (Table 6). Additionally, correct classification rates of validation sets varied from 97.34% to 98.40% (Table 6). The comparison parameters for SE, SP, MCC, and EFF (Tables 4 and 6), the results highlight classification abilities of data fusion OPLS-DA models were better than the individual data set models. What is more, the RS_OPLS-DA model was the optimum classification model when using low-level data fusion strategy (Tables 5 and 6).

Table 6. The major parameters of OPLS-DA models based on low-level data fusion strategy.

Model	Class	Calibration Set				Validation Set			
		I	II	III	IV	I	II	III	IV
RS_OPLS-DA	ACC (%)	99.46	100.00	99.46	98.92	97.87	98.94	97.87	98.94
	SE	1.00	1.00	1.00	0.96	0.96	1.00	0.97	0.96
	SP	0.99	1.00	0.99	1.00	0.99	0.99	0.98	1.00
	MCC	0.99	1.00	0.99	0.97	0.95	0.96	0.95	0.97
	EFF	1.00	1.00	1.00	0.98	0.97	0.99	0.98	0.98

Table 6. *Cont.*

RL_OPLS-DA	ACC (%)	99.46	100.00	100.00	99.46	95.74	97.87	97.87	97.87
	SE	1.00	1.00	1.00	0.98	0.88	1.00	0.97	0.96
	SP	0.99	1.00	1.00	1.00	0.99	0.97	0.98	0.99
	MCC	0.99	1.00	1.00	0.99	0.89	0.93	0.95	0.94
	EFF	1.00	1.00	1.00	0.99	0.93	0.99	0.98	0.97
SL_OPLS-DA	ACC (%)	100.00	100.00	100.00	100.00	94.68	98.94	97.87	97.87
	SE	1.00	1.00	1.00	1.00	1.00	0.94	0.93	0.91
	SP	1.00	1.00	1.00	1.00	0.93	1.00	1.00	1.00
	MCC	1.00	1.00	1.00	1.00	0.88	0.96	0.95	0.94
	EFF	1.00	1.00	1.00	1.00	0.96	0.97	0.96	0.96
RSL_OPLS-DA	ACC (%)	99.46	100.00	100.00	99.46	96.81	98.94	97.87	97.87
	SE	1.00	1.00	1.00	0.98	0.92	1.00	0.97	0.96
	SP	0.99	1.00	1.00	1.00	0.99	0.99	0.98	0.99
	MCC	0.99	1.00	1.00	0.99	0.92	0.96	0.95	0.94
	EFF	1.00	1.00	1.00	0.99	0.95	0.99	0.98	0.97

2.3.2. Mid-Level Data Fusion

At the end of the research, the feasibility for further optimizing the model parameters by feature subset selection and data fusion was investigated (Figure 11). Variables selection was one of the steps of the mid-level data fusion strategy. For the RF model, the "Boruta" algorithm was used to identify important chromatographic signal variables that significantly contributed to the classification performance. "Boruta" selection was finished based on three RM models that were built while using data sets of rhizomes (3839 variables), stems (4140 variables), and leaves (4140 variables), respectively. After comparing original attributes' importance with importance achievable at random, 200 variables of rhizome data set, 305 variables of stem data set, and 359 of variables for leaf data set were retained as relevant features variables for sample discrimination (Figures S9–S11). Subsequently, those feature subsets were combined as a new data block and the fused data set (505 variables for RS, 559 variables for RL, 664 variables for SL, and 864 variables for RSL) was used to establish final classification models. The optimum n_{tree} and m_{try} values of RS_RF, RL_RF, SL_RF, and RSL_RF model could be found in Figure S12.

Table 7 lists the statistical results for the classification ability of the four RF models based on mid-level data fusion. The average accuracies of the calibration set and validation set were achieved for 96.44% and 97.21% by using RF algorithm. It is notable that the RL_RF model had accuracies that ranged from 94.09% to 99.46% in the calibration set and accuracy ranging from 96.81% to 100% in the validation set. In addition, parameters of SE (0.87–1.00), SP (0.94–1.00), MCC (0.87–1.00), and EFF (0.92–1.00) for each class of RL_RF model were higher than most RF classification models. As a result, mid-level data fusion strategy could eliminate the unnecessary variables, enhance model classification ability, and improve the phenomenon of imbalance category recognition in the RF model relative to low-level data fusion strategy.

For the OPLS-DA model, in front of all, three independent classification models were built while using original data sets of rhizome, stem, and leaf, respectively. Subsequently, the VIP value of variables in different classification models was calculated by SIMCA software. The results showed (Figure S12) that a total of 4486 variables (1309 variables selected from rhizome data set, 1538 variables selected from stem data set and 1639 variables selected from leaf data set) VIP values were greater than 1. Those variables with large importance for the geographical traceability of samples were combined into a new data set (2847 variables for RS, 2948 variables for RL, 3177 variables for SL, and 4486 variables for RSL) for final classification model building. The R^2 and Q^2 values and the permutation plot of RS_OPLS-DA, RL_OPLS-DA, SL_OPLS-DA, and RSL_OPLS-DA model were shown in Table S2 and Figures S21–S24.

Table 7. The major parameters of RF models based on mid-level data fusion strategy.

Model	Class	Calibration Set				Validation Set			
		I	II	III	IV	I	II	III	IV
RS_RF	ACC (%)	99.46	99.46	94.09	95.16	98.94	100.00	96.81	97.87
	SE	0.98	0.97	0.95	0.87	1.00	1.00	0.97	0.91
	SP	1.00	1.00	0.94	0.98	0.99	1.00	0.97	1.00
	MCC	0.99	0.98	0.87	0.87	0.97	1.00	0.93	0.94
	EFF	0.99	0.98	0.94	0.92	0.99	1.00	0.97	0.96
RL_RF	ACC (%)	95.70	96.77	96.24	97.31	91.49	98.94	91.49	94.68
	SE	0.92	0.88	0.97	0.93	0.92	1.00	0.86	0.78
	SP	0.97	0.99	0.96	0.99	0.91	0.99	0.94	1.00
	MCC	0.89	0.88	0.91	0.93	0.80	0.96	0.80	0.86
	EFF	0.94	0.93	0.96	0.96	0.92	0.99	0.90	0.88
SL_RF	ACC (%)	95.16	96.77	93.55	96.24	97.87	100.00	98.94	96.81
	SE	0.94	0.81	0.95	0.89	0.96	1.00	1.00	0.91
	SP	0.96	1.00	0.93	0.99	0.99	1.00	0.98	0.99
	MCC	0.88	0.88	0.86	0.90	0.95	1.00	0.98	0.91
	EFF	0.95	0.90	0.94	0.94	0.97	1.00	0.99	0.95
RSL_RF	ACC (%)	97.85	99.46	94.09	95.70	96.81	100.00	95.74	98.94
	SE	0.94	0.97	0.93	0.91	0.96	1.00	0.93	0.96
	SP	0.99	1.00	0.95	0.97	0.97	1.00	0.97	1.00
	MCC	0.94	0.98	0.86	0.88	0.92	1.00	0.90	0.97
	EFF	0.97	0.98	0.94	0.94	0.97	1.00	0.95	0.98

The classification results showed that average accuracies of calibration and validation sets were achieved for 99.66% and 96.81%, respectively (Table 8). The four models exhibit good performances (MCC values ranged from 0.96 to 1.00 and EFF values ranged from 0.92 to 1.00 (Table 8). OPLS-DA models based on mid-level data fusion and low-level data fusion showed similar accuracy and model performance although feature selection was useful for reducing irrelevant variable when classifying samples.

Table 8. The major parameters of OPLS-DA models based on mid-level data fusion strategy.

Model	Class	Calibration Set				Validation Set			
		I	II	III	IV	I	II	III	IV
RS_OPLS-DA	ACC (%)	100.00	100.00	99.46	99.46	93.62	97.87	94.68	98.94
	SE	1.00	1.00	1.00	0.98	0.88	1.00	0.90	0.96
	SP	1.00	1.00	0.99	1.00	0.96	0.97	0.97	1.00
	MCC	1.00	1.00	0.99	0.99	0.84	0.93	0.87	0.97
	EFF	1.00	1.00	1.00	0.99	0.92	0.99	0.93	0.98
RL_OPLS-DA	ACC (%)	100.00	100.00	99.46	99.46	96.81	97.87	97.87	98.94
	SE	1.00	1.00	1.00	0.98	0.92	1.00	0.97	0.96
	SP	1.00	1.00	0.99	1.00	0.99	0.97	0.98	1.00
	MCC	1.00	1.00	0.99	0.99	0.92	0.93	0.95	0.97
	EFF	1.00	1.00	1.00	0.99	0.95	0.99	0.98	0.98
SL_OPLS-DA	ACC (%)	100.00	100.00	98.92	98.92	93.62	97.87	94.68	96.81
	SE	1.00	1.00	0.98	0.98	0.92	0.94	0.90	0.91
	SP	1.00	1.00	0.99	0.99	0.94	0.99	0.97	0.99
	MCC	1.00	1.00	0.97	0.97	0.85	0.92	0.87	0.91
	EFF	1.00	1.00	0.99	0.99	0.93	0.96	0.93	0.95
RSL_OPLS-DA	ACC (%)	100.00	100.00	99.46	99.46	95.74	98.94	96.81	97.87
	SE	1.00	1.00	1.00	0.98	0.92	1.00	0.93	0.96
	SP	1.00	1.00	0.99	1.00	0.97	0.99	0.98	0.99
	MCC	1.00	1.00	0.99	0.99	0.89	0.96	0.92	0.94
	EFF	1.00	1.00	1.00	0.99	0.95	0.99	0.96	0.97

Overall, it can be seen that there is an improvement in the results that were provided by data fusion when compared with performances of models based on independent data sets. When considering the similar accuracy and a higher SE, SP, MCC, and EFF values between calibration set and validation set, the RS_OPLS-DA models that were based on low-level data fusion strategy was the best performance.

3. Materials and Methods

3.1. Plant Material Collection

Plant materials (29 population and 280 individuals) of *G. rigescens* were collected in the fall of 2012 and 2013 at the time of local traditional harvest period, at the different location of Yunnan, Guizhou, and Sichuan (Figure 12). Four producing areas were divided according to the location of population. (I) low latitudes area, with latitudes ranging from 23.92–23.66° N, South of Yunnan (eight population and 76 individuals), (II) mid-latitude area, with latitudes ranges from 24.95–25.06° N, Middle of Yunnan (five population and 48 individuals), (III) mid-high latitude area, with latitudes ranges from 26.49–26.64° N, Northwest of Yunnan and West of Guizhou (nine population and 76 individuals 87), and (IV) high latitude area, with latitudes ranges from 27.34–28.52° N, Hengduan Mountains Region of Yunnan and mountainous regions of Southwest of Sichuan (seven population and 69 individuals). The fresh materials were authenticated and transported to the laboratory of Yuxi normal University. Subsequently, samples were wash cleaning and dried at 50 °C as soon as possible. At last, all samples (rhizomes, stems and leaves) were stored in a relatively dry environment prior to the extraction procedure.

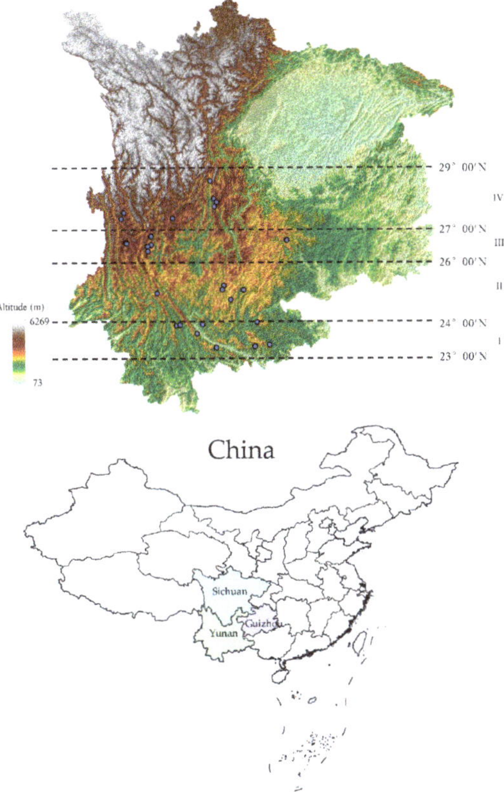

Figure 12. Geographical distribution of sample information.

3.2. Chemicals and Reagents

HPLC-grade acetonitrile, methanol (MeOH) were supplied by Thermo Fisher Scientific (Waltham, MA, USA). HPLC-grade formic acid was purchased from Sigma-Aldrich (Steinheim, Germany). Deionized water was obtained from Wahaha Group Co., Ltd. (Hangzhou, Zhejiang, China). The primary grade reference standards loganin (purity: ≥98%), 6′-O-β-D-glucopyranosylgentiopicroside (purity: ≥98%), swertiamarine (purity: ≥98%), gentiopicroside (purity: ≥98%), and sweroside (purity: ≥98%) were purchased from the Chinese National Institute for Food and Drug Control (Beijing, China), Shanghai Shifeng Biological Technology (Shanghai, China), respectively.

3.3. Sample Preparation

The dried samples (rhizomes, stems, and leaves) were ground and then passed through a 100 mesh sieves. Each sample powder (25 mg) was accurately weighed and extracted while using 1.5 mL 80% methanol-water solution, at 25 °C. The samples were extracted while using an Ultrasonic extractor for 40 min. The final extract was filtered with a 0.22 μm syringe filter into an HPLC vial and then subjected to HPLC analysis [16,58].

3.4. Instrumentation and HPLC Analysis

Chromatographic analyses were performed with an Agilent 1260 Infinity LC system (Agilent Technologies, Santa Clara, CA, USA), which was equipped with a G1315D diode-array detector, a G1329B ALS autosampler, and a thermostated column compartment. The HPLC fingerprint was recorded by Chemstation software (Agilent Technologies, Waldbron, Germany).

The analytical separation was adopted from a published method for chemical fingerprinting analysis [16]. The separation was achieved on a reversed phase C18 (Agilent Intersil, 5 μm, 4.6 × 150 mm) column (Agilent, Santa Clara, CA, USA). The composition of the mobile phase was: (A) 0.1% phosphoric acid in water and (B) 100% acetonitrile. The separation was as follows: 0.00–2.50 min: 7–10% B, 2.50–20.00 min: 10–26% B, 20.00–29.02 min: 26–58.3% B, 29.02–30.00 min: 58.3–90% B. The column was subsequently washed with 90% B and re-equilibrated with 7% B prior to injection of the next sample. The flow rate was 1.0 mL/min and the column temperature was 30 °C. The injection volume was 5 μL and the detective wavelength of UV spectra was set at 241 nm. Chromatographic data was processed while using OpenLab software (Agilent Technologies) [16,58].

3.5. Data Analysis

HPLC fingerprints from the 280 rhizome samples, 280 stem samples, and 280 leaf samples, a total of 840 fingerprint data was exported in CSV format and imported to MATLAB R2018b (The MathWorks, Inc., Natick, MA, USA), which was used for correlation optimized warping (COW) alignment preprocessing of chromatographic fingerprint. MATLAB code of COW is freely available from www.models.kvl.dk. The preprocessing fingerprint was analyzed in the following work [59].

Exploratory data analysis (EDA) is necessary for building predictive models [60,61]. It can help in determining interesting correlations among all of the samples or variables and summarize data sets main characteristics [60]. Principal component analysis (PCA) is a popular primary tool in EDA [61,62]. It is often used to visualize the relatedness between samples and explains the variance in the data. Hence, PCA, as an unsupervised pattern recognition technique, was widely used to extract key information from chemical fingerprint for geographical origin or Modelling Research [61].

Unlike PCA, orthogonal partial least squares discriminant analysis (OPLS-DA) is a supervised pattern recognition technique. As an extension of PLS, an inbuilt orthogonal signal correction filter was incorporated in the OPLS-DA model [56]. This algorithm effectively divides the X variable into two parts: one part that is related to class information (Y-predictive) and the other is orthogonal or unrelated to class information (Y-uncorrelated). Therefore, interpretability and prediction performance of the model was enhanced [56].

Random forest (RF) is another supervised pattern recognition technique utilized in the study. RF is an ensemble learning method [55]. A large number of trees were produced by RF algorithm in order to improve model predictive ability, and trees' decision results were combined as final decision results. In other words, the more trees built in the random forest classifier, the higher accuracy could be achieved. However, many researches showed that an optimum tree number was of great importance in modeling classification performance [33,46].

In this work, exploratory data analysis of HPLC fingerprints of *G. rigescens* grown in four different latitudes was finished with PCA. Two supervised pattern recognition techniques, OPLS-DA and RF, were applied to build classification models for *G. rigescens* producing areas. SIMCA 14.1 software managed PCA and OPLS-DA (Umetrics AB, Umea, Sweden). RF classification models were established with R 3.5.1 program and package randomForest (Version 4.6-14) [63].

Data Fusion Strategy

In the case of low-level fusion strategy (Figure 11), different subsets HPLC fingerprint data matrix of rhizomes, stems, and leaves) are straightforwardly concatenated and compiled into a new chromatographic data matrix for subsequent classification model construction [45,46]. Furthermore, each subset must be totally aligned and keep all the variables on the same scale before subsets reconnection [45,46].

In the case of mid-level fusion (Figure 11), the first step of data treatment is feature selection that is based on rhizomes, stems, or leaves classification models. When compared with the raw data sets, feature selection of subsets minimizes the data content and reduces data dimensions. Subsequently, new subsets of rhizomes, stems, and leaves were rebuilt while using variables of feature selection [45]. At last, those subsets are concatenated and compiled into a final data matrix for model construction [45].

In the research, relevant variables of RF classification models were determined by the R software package Boruta [64], and VIP was used for important variables selection of OPLS-DA [65].

3.6. Model Evaluation

Five parameters, including accuracy (ACC), sensitivity (SE), specificity (SP), efficiency (EFF), and Matthews correlation coefficient (MCC) were applied to evaluate the identification ability of RF and OPLS-DA models. The ruggedness of OPLS-DA model was investigated through 200 times permutation tests. Furthermore, cumulative prediction ability (Q^2), cumulative interpretation ability (R^2), root mean square error of estimation (RMSEE), root mean square error of cross-validation (RMSECV), and root mean square error of prediction (RMSEP) were important evaluation indexes for the predictive power of OPLS-DA model [33,66].

Values of TP (Correctly identified samples of positive class), TN (correctly identified samples of negative class), FN (incorrectly identified samples of positive class), and FP (incorrectly identified samples of negative class) were calculated according to confusion matrixes of classification models. Subsequently, ACC, SE, SP, EFF, and MCC were calculated while using Equations (1)–(5) and values of Q^2, R^2, RMSEE, RMSECV, and RMSEP computed by software SIMCA 14.1.

$$\text{ACC} = \frac{(\text{TN} + \text{TP})}{(\text{TP} + \text{TN} + \text{FP} + \text{FN})} \tag{1}$$

$$\text{SE} = \frac{\text{TP}}{(\text{TP} + \text{FN})} \tag{2}$$

$$\text{SP} = \frac{\text{TN}}{(\text{TN} + \text{FP})} \tag{3}$$

$$\text{EFF} = \sqrt{\text{SE} \times \text{SP}} \tag{4}$$

$$\text{MCC} = \frac{(\text{TP} \times \text{TN} - \text{FP} \times \text{FN})}{\sqrt{(\text{TP} + \text{FP})(\text{TP} + \text{FN})(\text{TN} + \text{FP})(\text{TN} + \text{FN})}} \tag{5}$$

For model performance, lower values of RMSEE, RMSECV, and RMSEP mean better predictive ability for the models. Conversely, the closer that values of ACC, SE, SP, EFF, MCC, and Q^2, R^2 are to 1, the more well performance the model is.

4. Conclusions

The findings in this study showed that *G. rigescens* chemical profiles were influenced by the latitude gradients of producing areas and lower latitudes and higher latitudes samples seemed to be clearly distinguishable. According to the score plots of PCA and OPLS-DA, the phytochemical geographic variation of the overground and underground part along the latitude gradients was visualized. Subsequently, the potential of fingerprint data obtained while using HPLC-DAD to discriminate and classify *G. rigescens* grown in four different latitudes was investigated. Additionally, RF and OPLS-DA models were used to develop an effective way for geographical traceability of the *G. rigescens* that were grown in four different latitudes. When using independent data sets to build models, rhizomes data set combined with OPLS-DA presented the best performance with a classification accuracy of calibration and validation set varied from 94.68% to 98.94%. In a further step, the feasibility of combining the chromatographic fingerprint data from overground and underground organs was investigated based on two kinds of data fusion strategies in order to improve the performance of classification models: low-level and mid-level. Notably, classification performances of OPLS-DA models were efficiently improved by low-level data fusion strategy and better performances of RF models appeared to be achieved by mid-level data fusion strategy. Although satisfactory results were obtained with both RF and OPLS-DA based on two kinds of data fusion strategies, OPLS-DA combined with rhizome-stem fusion data set was the optimum model for discriminating *G. rigescens* samples according to their grown latitudes, with an accuracy of (97.87–100.00%), SE of (0.96–1.00), SP of (0.98–1.00), MCC of (0.95–1.00), and EFF of (0.97–1.00).

Supplementary Materials: The following are available online. Figure S1: Variation of stems score plots along the latitude gradients, Figure S2: Variation of stems score plots between the adjacent latitudes, Figure S3: Variation of leaves score plots along the latitude gradients, Figure S4: Variation of leaves score plots between the adjacent latitudes, Figure S5: Permutation plot of the OPLS-DA of rhizome samples, Figure S6: Permutation plot of the OPLS-DA of stem samples, Figure S7: Permutation plot of the OPLS-DA of leaf samples, Figure S8: The n_{tree} and m_{try} screening of RF models based on low-level data fusion strategy, Figure S9: Result of variables selection of rhizome fingerprint data based on "Boruta" algorithm, Figure S10. Result of variables selection of stem fingerprint data based on "Boruta" algorithm, Figure S11: Result of variables selection of leaf fingerprint data based on "Boruta" algorithm, Figure S12: The n_{tree} and m_{try} screening of RF models based on mid-level data fusion strategy, Figure S13: The importance variables of OPLS-DA models of rhizomes, stems and leaves fingerprints data, Figure S14: Permutation testing (200 times) of the R_OPLS-DA model, Figure S15: Permutation testing (200 times) of the S_OPLS-DA model, Figure S16: Permutation testing (200 times) of the L_OPLS-DA model, Figure S17: Permutation testing (200 times) of the RS_OPLS-DA model based on low-level data fusion, Figure S18: Permutation testing (200 times) of the RL_OPLS-DA model based on low-level data fusion, Figure S19: Permutation testing (200 times) of the SL_OPLS-DA model based on low-level data fusion, Figure S20: Permutation testing (200 times) of the RSL_OPLS-DA model based on low-level data fusion, Figure S21: Permutation testing (200 times) of the RS_OPLS-DA model based on mid-level data fusion, Figure S22: Permutation testing (200 times) of the RL_OPLS-DA model based on mid-level data fusion, Figure S23: Permutation testing (200 times) of the SL_OPLS-DA model based on mid-level data fusion, Figure S24: Permutation testing (200 times) of the RSL_OPLS-DA model based on mid-level data fusion, Table S1: The evaluation indexes for predictive power of OPLS-DA model of rhizome, stem and leaf, Table S2: The evaluation indexes for predictive power of OPLS-DA models based on low-level and mid-level data fusion strategies.

Author Contributions: H.Y. and Y.-Z.W. designed the project and revised the manuscript. T.S. performed the experiments, analyzed the data and wrote the manuscript.

Funding: This research was supported by the Key Project of Yunnan Provincial Natural Science Foundation (2017FA049), the Projects for Applied Basic Research in Yunnan (2017FH001-028), Biodiversity Survey, Monitoring and Assessment (2019HB2096001006) and the Department of Science and Technology of Yunnan Province (2018IA075).

Conflicts of Interest: The authors declare no conflict of interest.

References

1. Flora of China Editorial Committee. *Flora of China*; Science Press and Missouri Botanical Garden Press: Beijing, China, 1995; Volume 22.
2. Mustafa, A.M.; Caprioli, G.; Ricciutelli, M.; Maggi, F.; Marín, R.; Vittori, S.; Sagratini, G. Comparative HPLC/ESI-MS and HPLC/DAD study of different populations of cultivated, wild and commercial *Gentiana lutea* L. *Food Chem.* **2015**, *174*, 426–433. [CrossRef] [PubMed]
3. Pan, Y.; Zhao, Y.L.; Zhang, J.; Li, W.Y.; Wang, Y.Z. Phytochemistry and pharmacological activities of the genus *Gentiana* (Gentianaceae). *Chem. Biodivers.* **2016**, *13*, 107–150. [CrossRef] [PubMed]
4. Jiang, R.W.; Wong, K.L.; Chan, Y.M.; Xu, H.X.; But, P.P.H.; Shaw, P.C. Isolation of iridoid and secoiridoid glycosides and comparative study on *Radix gentianae* and related adulterants by HPLC analysis. *Phytochemistry* **2005**, *66*, 2674–2680. [CrossRef] [PubMed]
5. Xu, Y.; Li, Y.; Maffucci, K.; Huang, L.; Zeng, R. Analytical methods of phytochemicals from the genus *Gentiana*. *Molecules* **2017**, *22*, 2080. [CrossRef] [PubMed]
6. Mustafa, A.M.; Ricciutelli, M.; Maggi, F.; Sagratini, G.; Vittori, S.; Caprioli, G. Simultaneous determination of 18 bioactive compounds in Italian bitter liqueurs by reversed-phase high-performance liquid chromatography—Diode array detection. *Food Anal. Method* **2014**, *7*, 697–705. [CrossRef]
7. Mirzaee, F.; Hosseini, A.; Jouybari, H.B.; Davoodi, A.; Azadbakht, M. Medicinal, biological and phytochemical properties of *Gentiana* species. *J. Tradit. Complement. Med.* **2017**, *7*, 400–408. [CrossRef] [PubMed]
8. Gao, L.J.; Li, J.Y.; Qi, J.H. Gentisides A and B, two new neuritogenic compounds from the traditional Chinese medicine *Gentiana rigescens* franch. *Bioorgan. Med. Chem.* **2010**, *18*, 2131–2134. [CrossRef]
9. Gao, L.J.; Xiang, L.; Luo, Y.; Wang, G.F.; Li, J.Y.; Qi, J.H. Gentisides C-K: Nine new neuritogenic compounds from the traditional Chinese medicine *Gentiana rigescens* franch. *Bioorgan. Med. Chem.* **2010**, *18*, 6995–7000. [CrossRef]
10. Li, J.; Gao, L.J.; Sun, K.Y.; Xiao, D.; Li, W.Y.; Xiang, L.; Qi, J.H. Benzoate fraction from *Gentiana rigescens* franch alleviates scopolamine-induced impaired memory in mice model in vivo. *J. Ethnopharmacol.* **2016**, *193*, 107–116. [CrossRef]
11. Mustafa, A.M.; Caprioli, G.; Dikmen, M.; Kaya, E.; Maggi, F.; Sagratini, G.; Vittori, S.; öztürk, Y. Evaluation of neuritogenic activity of cultivated, wild and commercial roots of *Gentiana lutea* L. *J. Funct. Foods* **2015**, *19*, 164–173. [CrossRef]
12. Pharmacopoeia, C.C. *Pharmacopoeia of the People's Republic of China*; China Medicinal Science Press: Beijing, China, 2015.
13. Pan, Y.; Zhang, J.; Shen, T.; Zhao, Y.L.; Zuo, Z.T.; Wang, Y.Z.; Li, W.Y. Investigation of chemical diversity in different parts and origins of ethnomedicine *Gentiana rigescens* franch using targeted metabolite profiling and multivariate statistical analysis. *Biomed. Chromatogr.* **2016**, *30*, 232–240. [CrossRef] [PubMed]
14. Li, J.; Zhang, J.; Zhao, Y.L.; Huang, H.Y.; Wang, Y.Z. Comprehensive quality assessment based specific chemical profiles for geographic and tissue variation in *Gentiana rigescens* using HPLC and FTIR method combined with principal component analysis. *Front. Chem.* **2017**, *5*, 125. [CrossRef] [PubMed]
15. Wu, Z.; Zhao, Y.L.; Zhang, J.; Wang, Y.Z. Quality assessment of *Gentiana rigescens* from different geographical origins using FT-IR spectroscopy combined with HPLC. *Molecules* **2017**, *22*, 1238. [CrossRef] [PubMed]
16. Wang, Y.; Shen, T.; Zhang, J.; Huang, H.Y.; Wang, Y.Z. Geographical authentication of *Gentiana rigescens* by high-performance liquid chromatography and infrared spectroscopy. *Anal. Lett.* **2018**, *51*, 2173–2191. [CrossRef]
17. Ailer, B.; Avramov, S.; Banjanac, T.; Cvetković, J.; Nestorović Živković, J.; Patenković, A.; Mišić, D. Secoiridoid glycosides as a marker system in chemical variability estimation and chemotype assignment of *Centaurium erythraea* Rafn from the Balkan Peninsula. *Ind. Crop. Prod.* **2012**, *40*, 336–344.
18. Yang, Y.M.; Tian, K.; Hao, J.M.; Pei, S.J.; Yang, Y.X. Biodiversity and biodiversity conservation in Yunnan, China. *Biodivers. Conserv.* **2004**, *13*, 813–826. [CrossRef]
19. Fan, Z.X.; Thomas, A. Spatiotemporal variability of reference evapotranspiration and its contributing climatic factors in Yunnan Province, SW China, 1961–2004. *Clim. Chang.* **2013**, *116*, 309–325. [CrossRef]
20. Liu, M.X.; Xu, X.L.; Sun, A.Y.; Wang, K.L.; Yue, Y.M.; Tong, X.W.; Liu, W. Evaluation of high-resolution satellite rainfall products using rain gauge data over complex terrain in southwest China. *Theor. Appl. Climatol.* **2015**, *119*, 203–219. [CrossRef]

21. Tang, Q.H.; Ge, Q.S. *Atlas of Environmental Risks Facing China under Climate Change*; Springer Verlag: Berlin, Germany, 2018.
22. Zhao, Z.Z.; Guo, P.; Brand, E. The formation of daodi medicinal materials. *J. Ethnopharmacol.* **2012**, *140*, 476–481. [CrossRef]
23. Sun, M.M.; Li, L.; Wang, M.; van Wijk, E.; He, M.; van Wijk, R.; Koval, S.; Hankemeier, T.; van der Greef, J.; Wei, S.L. Effects of growth altitude on chemical constituents and delayed luminescence properties in medicinal rhubarb. *J. Photoch. Photobio. B* **2016**, *162*, 24–33. [CrossRef]
24. Song, X.Y.; Jin, L.; Shi, Y.P.; Li, Y.D.; Chen, J. Multivariate statistical analysis based on a chromatographic fingerprint for the evaluation of important environmental factors that affect the quality of *Angelica sinensis*. *Anal. Methods* **2014**, *6*, 8268–8276. [CrossRef]
25. Yao, R.Y.; Heinrich, M.; Zou, Y.F.; Reich, E.; Zhang, X.L.; Chen, Y.; Weckerle, C.S. Quality variation of Goji (fruits of *Lycium* spp.) in China: A comparative morphological and metabolomic analysis. *Front. Pharmacol.* **2018**, *9*, 151. [CrossRef] [PubMed]
26. Huang, Y.P.; Wu, Z.W.; Su, R.H.; Ruan, G.H.; Du, F.Y.; Li, G.K. Current application of chemometrics in traditional Chinese herbal medicine research. *J. Chromatogr. B* **2016**, *1026*, 27–35. [CrossRef] [PubMed]
27. Zhang, C.; Zheng, X.; Ni, H.; Li, P.; Li, H.J. Discovery of quality control markers from traditional Chinese medicines by fingerprint-efficacy modeling: Current status and future perspectives. *J. Pharmaceut. Biomed.* **2018**, *159*, 296–304. [CrossRef] [PubMed]
28. Chen, D.D.; Xie, X.F.; Ao, H.; Liu, J.L.; Peng, C. Raman spectroscopy in quality control of Chinese herbal medicine. *J. Chin. Med. Assoc.* **2017**, *80*, 288–296. [CrossRef] [PubMed]
29. Wang, P.; Yu, Z.G. Species authentication and geographical origin discrimination of herbal medicines by near infrared spectroscopy: A review. *J. Pharm. Anal.* **2015**, *5*, 277–284. [CrossRef] [PubMed]
30. Qi, L.M.; Zhang, J.; Zhao, Y.L.; Zuo, Z.T.; Wang, Y.Z.; Jin, H. Characterization of *Gentiana rigescens* by ultraviolet-visible and infrared spectroscopies with chemometrics. *Anal. Lett.* **2017**, *50*, 1497–1511. [CrossRef]
31. Zhao, Y.L.; Zhang, J.; Jin, H.; Zhang, J.Y.; Shen, T.; Wang, Y.Z. Discrimination of *Gentiana rigescens* from different origins by fourier transform infrared spectroscopy combined with chemometric methods. *J. Aoac Int.* **2015**, *98*, 22–26. [CrossRef]
32. Lee, D.Y.; Kang, K.B.; Kim, J.; Kim, H.J.; Sung, S.H. Classficiation of bupleuri radix according to geographical origins using near infrared spectroscopy (NIRS) combined with supervised pattern recognition. *Nat. Prod. Sci.* **2018**, *24*, 164. [CrossRef]
33. Pei, Y.F.; Wu, L.H.; Zhang, Q.Z.; Wang, Y.Z. Geographical traceability of cultivated *Paris polyphylla* var.*yunnanensis* using ATR-FTMIR spectroscopy with three mathematical algorithms. *Anal. Methods* **2019**, *11*, 113–122. [CrossRef]
34. Chen, H.; Lin, Z.; Tan, C. Fast discrimination of the geographical origins of notoginseng by near-infrared spectroscopy and chemometrics. *J. Pharmaceut. Biomed.* **2018**, *161*, 239–245. [CrossRef] [PubMed]
35. Wang, Q.Q.; Huang, H.Y.; Wang, Y.Z. Geographical authentication of *Macrohyporia cocos* by a data fusion method combining ultra-fast liquid chromatography and fourier transform infrared spectroscopy. *Molecules* **2019**, *24*, 1320. [CrossRef] [PubMed]
36. Ma, X.D.; Fan, Y.X.; Jin, C.C.; Wang, F.; Xin, G.Z.; Li, P.; Li, H.J. Specific targeted quantification combined with non-targeted metabolite profiling for quality evaluation of *Gastrodia elata* tubers from different geographical origins and cultivars. *J. Chromatogr. A* **2016**, *1450*, 53–63. [CrossRef] [PubMed]
37. Tang, J.F.; Li, W.X.; Zhang, F.; Li, Y.H.; Cao, Y.J.; Zhao, Y.; Li, X.L.; Ma, Z.J. Discrimination of radix polygoni multiflori from different geographical areas by UPLC-QTOF/MS combined with chemometrics. *Chin. Med.* **2017**, *12*, 1–12. [CrossRef] [PubMed]
38. Zhu, L.X.; Xu, J.; Wang, R.J.; Li, H.X.; Tan, Y.Z.; Chen, H.B.; Dong, X.P.; Zhao, Z.Z. Correlation between quality and geographical origins of *Poria cocos* revealed by qualitative fingerprint profiling and quantitative determination of triterpenoid acids. *Molecules* **2018**, *23*, 2200. [CrossRef] [PubMed]
39. Sun, L.L.; Wang, M.; Zhang, H.J.; Liu, Y.N.; Ren, X.L.; Deng, Y.R.; Qi, A.D. Comprehensive analysis of polygoni multiflori radix of different geographical origins using ultra-high-performance liquid chromatography fingerprints and multivariate chemometric methods. *J. Food Drug Anal.* **2018**, *26*, 90–99. [CrossRef] [PubMed]
40. Cuadros-Rodríguez, L.; Ruiz-Samblás, C.; Valverde-Som, L.; Pérez-Castaño, E.; González-Casado, A. Chromatographic fingerprinting: An innovative approach for food 'identitation' and food authentication—A tutorial. *Anal. Chim. Acta* **2016**, *909*, 9–23. [CrossRef] [PubMed]

41. Lucio-Gutiérrez, J.R.; Coello, J.; Maspoch, S. Enhanced chromatographic fingerprinting of herb materials by multi-wavelength selection and chemometrics. *Anal. Chim. Acta* **2012**, *710*, 40–49. [CrossRef]
42. Zhang, L.L.; Liu, Y.Y.; Liu, Z.L.; Wang, C.; Song, Z.Q.; Liu, Y.X.; Dong, Y.Z.; Ning, Z.C.; Lu, A.P. Comparison of the roots of *Salvia miltiorrhiza* bunge (danshen) and its variety *S. miltiorrhiza* Bge f. Alba (baihua danshen) based on multi-wavelength HPLC-fingerprinting and contents of nine active components. *Anal. Methods* **2016**, *8*, 3171–3182. [CrossRef]
43. Wang, X.; Li, B.Q.; Xu, M.L.; Liu, J.J.; Zhai, H.L. Quality assessment of traditional Chinese medicine using HPLC-PAD combined with tchebichef image moments. *J. Chromatogr. B* **2017**, *1040*, 8–13. [CrossRef]
44. Jiménez-Carvelo, A.M.; Cruz, C.M.; Olivieri, A.C.; González-Casado, A.; Cuadros-Rodríguez, L. Classification of olive oils according to their cultivars based on second-order data using LC-DAD. *Talanta* **2019**, *195*, 69–76. [CrossRef] [PubMed]
45. Borràs, E.; Ferré, J.; Boqué, R.; Mestres, M.; Aceña, L.; Busto, O. Data fusion methodologies for food and beverage authentication and quality assessment—A review. *Anal. Chim. Acta* **2015**, *891*, 1–14. [CrossRef] [PubMed]
46. Li, Y.; Zhang, J.Y.; Wang, Y.Z. FT-MIR and NIR spectral data fusion: A synergetic strategy for the geographical traceability of *Panax notoginseng*. *Anal. Bioanal. Chem.* **2018**, *410*, 91–103. [CrossRef] [PubMed]
47. Wu, X.M.; Zuo, Z.T.; Zhang, Q.Z.; Wang, Y.Z. FT-MIR and UV–vis data fusion strategy for origins discrimination of wild *Paris polyphylla* Smith var. *yunnanensis*. *Vib. Spectrosc.* **2018**, *96*, 125–136. [CrossRef]
48. Wang, H.Y.; Song, C.; Sha, M.; Liu, J.; Li, L.P.; Zhang, Z.Y. Discrimination of medicine radix astragali from different geographic origins using multiple spectroscopies combined with data fusion methods. *J. Appl. Spectrosc.* **2018**, *85*, 313–319. [CrossRef]
49. Pei, Y.F.; Zhang, Q.Z.; Zuo, Z.T.; Wang, Y.Z. Comparison and Identification for rhizomes and leaves of *Paris yunnanensis* based on Fourier transform mid-Infrared spectroscopy combined with chemometrics. *Molecules* **2018**, *23*, 3343. [CrossRef] [PubMed]
50. Yang, H.; Liu, J.; Chen, S.; Hu, F.; Zhou, D. Spatial variation profiling of four phytochemical constituents in *Gentiana straminea* (Gentianaceae). *J. Nat. Med.* **2014**, *68*, 38–45. [CrossRef]
51. Lei, M.; Yu, X.H.; Li, M.; Zhu, W.X. Geographic origin identification of coal using near-infrared spectroscopy combined with improved random forest method. *Infrared Phys. Technol.* **2018**, *92*, 177–182. [CrossRef]
52. Sayago, A.; González-Domínguez, R.; Beltrán, R.; Fernández-Recamales, Á. Combination of complementary data mining methods for geographical characterization of extra virgin olive oils based on mineral composition. *Food Chem.* **2018**, *261*, 42–50. [CrossRef]
53. Jandrić, Z.; Haughey, S.A.; Frew, R.D.; Mccomb, K.; Galvin-King, P.; Elliott, C.T.; Cannavan, A. Discrimination of honey of different floral origins by a combination of various chemical parameters. *Food Chem.* **2015**, *189*, 52–59. [CrossRef]
54. Jolayemi, O.S.; Ajatta, M.A.; Adegeye, A.A. Geographical discrimination of palm oils (*Elaeis guineensis*) using quality characteristics and UV-visible spectroscopy. *Food Sci. Nutr.* **2018**, *6*, 773–782. [CrossRef] [PubMed]
55. Breiman, L. Random forests. *Mach. Learn.* **2001**, *45*, 5–32. [CrossRef]
56. Wold, S.; Antti, H.; Lindgren, F.; öhman, J. Orthogonal signal correction of near-infrared spectra. *Chemometr. Intell. Lab.* **1998**, *44*, 175–185. [CrossRef]
57. Boccard, J.L.; Rutledge, D.N. A consensus orthogonal partial least squares discriminant analysis (OPLS-DA) strategy for multiblock omics data fusion. *Anal. Chim. Acta* **2013**, *769*, 30–39. [CrossRef] [PubMed]
58. Lyv, W.Q.; Zhang, J.; Zuo, Z.T.; Wang, Y.Z.; Zhang, Q.Z. Quality evaluation of *Gentiana rigescence* by grey relational analysis method. *Chin. J. Exp. Tradit. Med. Formul.* **2017**, *23*, 66–73.
59. Skov, T.; van den Berg, F.; Tomasi, G.; Bro, R. Automated alignment of chromatographic data. *J. Chemometr.* **2006**, *20*, 484–497. [CrossRef]
60. Kürzl, H. Exploratory data analysis: Recent advances for the interpretation of geochemical data. *J. Geochem. Explor.* **1988**, *30*, 309–322. [CrossRef]
61. Esteki, M.; Shahsavari, Z.; Simal-Gandara, J. Food identification by high performance liquid chromatography fingerprinting and mathematical processing. *Food Res. Int.* **2019**, *122*, 303–317. [CrossRef]
62. Berrueta, L.A.; Alonso-Salces, R.M.; Héberger, K. Supervised pattern recognition in food analysis. *J. Chromatogr. A* **2007**, *1158*, 196–214. [CrossRef]
63. Liaw, A.; Wiener, M. Classification and regression by randomForest. *R News* **2002**, *2*, 18–22.

64. Kursa, M.B.; Rudnicki, W.R. Feature selection with the boruta package. *J. Stat. Softw.* **2010**, *36*, 1–13. [CrossRef]
65. Mehmood, T.; Liland, K.H.; Snipen, L.; Sæbø, S. A review of variable selection methods in partial least squares regression. *Chemometr. Intell. Lab.* **2012**, *118*, 62–69. [CrossRef]
66. Cao, D.S.; Hu, Q.N.; Xu, Q.S.; Yang, Y.N.; Zhao, J.C.; Lu, H.M.; Zhang, L.X.; Liang, Y.Z. In silico classification of human maximum recommended daily dose based on modified random forest and substructure fingerprint. *Anal. Chim. Acta* **2011**, *692*, 50–56. [CrossRef] [PubMed]

Sample Availability: Samples of the compound are not available from the authors.

Article

Screening 89 Pesticides in Fishery Drugs by Ultrahigh Performance Liquid Chromatography Tandem Quadrupole-Orbitrap Mass Spectrometer

Shou-Ying Wang [1,2], Cong Kong [1,3,*], Qing-Ping Chen [1,2] and Hui-Juan Yu [1,3,*]

1 Laboratory of Quality & Safety Risk Assessment for Aquatic products (Shanghai), Ministry of Agriculture and Rural Affairs, East China Sea Fisheries Research Institute, Shanghai 200090, China; magnolia7319@163.com (S.-Y.W.); chenqp128@163.com (Q.-P.C.)

2 College of Food Science & Technology, Shanghai Ocean University, Shanghai 201306, China

3 Key Laboratory of East China Sea Fishery Resources Exploitation, Ministry of Agriculture and Rural Affairs, East China Sea Fisheries Research Institute, Chinese Academy of Fishery Sciences, Shanghai 200090, China

* Correspondence: kongcong@gmail.com (C.K.); xdyh-7@163.com (H.-J.Y.)

Academic Editors: Marcello Locatelli, Angela Tartaglia, Dora Melucci, Abuzar Kabir, Halil Ibrahim Ulusoy, Victoria Samanidou, Evagelos Gikas and Patrick Chaimbault

Received: 5 July 2019; Accepted: 12 September 2019; Published: 17 September 2019

Abstract: Multiclass screening of drugs with high resolution mass spectrometry is of great interest due to its high time-efficiency and excellent accuracy. A high-scale, fast screening method for pesticides in fishery drugs was established based on ultrahigh performance liquid chromatography tandem quadrupole-Orbitrap high-resolution mass spectrometer. The target compounds - were diluted in methanol and extracted by ultrasonic treatment, and the extracts were diluted with MeOH-water (1:1, *v*/*v*) and centrifuged to remove impurities. The chromatographic separation was performed on an Accucore aQ-MS column (100 mm × 2.1 mm, 2.6 μm) with gradient elution using 0.1% formic acid in water (containing 5 mmol/L ammonium formate) and 0.1% formic acid in methanol (containing 5 mmol/L ammonium formate) in Full Scan/dd-MS2 (TopN) scan mode. A screening database, including mass spectrometric and chromatographic information, was established for identification of compounds. The screening detection limits of methods ranged between 1–500 mg/kg, the recoveries of real samples spiked with the concentration of 10 mg/kg and 100 mg/kg standard mixture ranged from 70% to 110% for more than sixty compounds, and the relative standard deviations (RSDs) were less than 20%. The application of this method showed that target pesticides were screened out in 10 samples out of 21 practical samples, in which the banned pesticide chlorpyrifos were detected in 3 out of the 10 samples.

Keywords: fishery drugs; high-resolution orbitrap mass spectrometry; pesticide; screening

1. Introduction

Aquaculture is estimated to provide half of aquatic products by 2030 from the farming of freshwater or marine areas [1]. There is inevitably going to be a need for intensive aquaculture developed to supply more products from this industry. According to the "Green food—fishery medicine application guideline (NY/T 755-2013)" in the Agricultural Industry Standards of the People's Republic of China [2], fishery medicine refers to the substances that prevent or treat diseases in aquaculture animals or purposefully regulate the physiology of animals, including chemicals, antibiotics, Chinese herbal medicines and biological products. It is also known as chemical inputs or veterinary medicinal products (VMPs) applied in aquaculture in Europe and the United States [3,4]. Chemical inputs from aquaculture include antifoulants, antibiotics, parasiticides, anesthetics and disinfectants [5], while parasiticides in fishery mainly contain avermectins, pyrethroids, hydrogen peroxide, and organophosphates [5,6].

Based on the Guidelines, ten kinds of fishery drugs originated from pesticide have been banned for aquatic animals and plants. However, the illegal or excessive addition of pesticides in the fishery drug, as well as uncontrollable and uncertain administration during culture process can lead to the accumulation and residue of these pesticides in aquatic product. Illegal and unregulated use of pesticide may occur in many aquaculture areas, and further threaten the food safety for human health. To protect the quality and safety of aquatic products, as well as the sustainable ecosystem, surveillance of pesticides components in fishery drugs should be conducted.

Ultrahigh performance liquid chromatography coupled to high-resolution mass spectrometry (UHPLC-HRMS) is a promising strategy for multi-component screening of pesticides [7–9]. HRMS could record full scan of the precursors or fragmented ions with high-resolution, as well as the relative isotopic abundance, and is virtually able to distinguish unlimited number of compounds from one set of analyzed data [10,11]. In the past, the chromatography coupled to Time of Flight Mass Spectrometry (ToF-MS) was used in the development of multiclass components screening methods [9,12,13]. However, comparing to ToF-MS, the orbitrap mass spectrometer can fast scan and simultaneously switch between positive and negative acquisition modes if there's no need to change mobile phase of chromatography unit [14,15]. The combination of quadrupole and Orbitrap for high-resolution mass spectrometry can acquire data with high throughput, excellent accuracy and better sensitivity, which provides an ideal platform for multiclass risk compound screening [16]. Therefore, more methods of screening detection with Orbitrap MS were developed. With this instrument, the data-dependent data acquisition mode scans the full mass distribution of all precursors and then selectively fragments them sequentially for secondary mass scanning according to their abundance. This scan mode allows the quantification of compounds with precursor ion abundancy and identification with corresponding fragment ions [17]. Moreover, due to the stable and high-resolution mass spectrum recorded at standard data provide enough dependency, the identification of targeted compound can be conducted by comparing their database rather than practically acquire data for standards every time [18,19].

In previous studies, the analysis of 139 pesticide residues in fruit and vegetable commodities was established based on the Q-Orbitrap MS, allowing the retrospective analysis of the data feature which cannot be achieved with QqQ [17]. Jia et al. have developed an untargeted screening method for 137 veterinary drugs and their metabolites (16 categories) in tilapia using UHPLC-Orbitrap MS [20]. Turnipseed established a wide-scope screening method for 70 veterinary drugs in fish, shrimp and eel using LC-Orbitrap MS [7]. Recently, a non-target data acquisition for target analysis workflow based on UHPLC/ESI Q-Orbitrap was examined for its performance in screening pesticide residues in fruit and vegetables [21]. However, there is a lack of works on the multi-component screening detection in fishery drugs, especially for pesticide component screening. A fast screening method for a wide range of pesticides detection can be preferred, as much more reagent, time, and labor can be saved to detect more harmful components for safety evaluation.

Our study aims to develop a more generic screening method for a wider scope of pesticides with a self-built database, which can keep the advantages of robustness, simplicity, and time-efficiency. In the current work, we investigated 89 possible pesticides that can be used in fishery-related industry and remained in aquatic products. The chromatographic and high-resolution mass spectra for these compounds were acquired with a UHPLC-quadrupole-Orbitrap HRMS after optimizing parameters. The useful fragment ions with high-resolution were explored and selected. Then, a database including the retention time, isotope pattern, ionization mode and adduct, characteristic fragment ions, was established. Identification rules for data comparison with real samples were also investigated. Finally, a fast pesticide screening method for fishery drug was developed in combination with a rapid pretreatment.

2. Results and Discussion

2.1. Full MS-ddMS^2 Scan for Identification and Qualification

Full MS-ddMS^2 detection mode was applied on UHPLC-ESI-Q-Orbitrap HRMS system, which is a different data acquisition from single (multiple) reaction monitoring on triple quadrupole mass spectrometry. The Orbitrap analyzer collected accurate mass of all precursor ions as the first identification step of compounds. The precursors of high abundance were isolated through quadrupole in the next round scanning. Each of the precursors can be fragmented sequentially in the HCD multipole, re-collected in C-trap, and analyzed through Orbitrap mass spectrometer. It should be noted that the accurate mass of precursors instead of their fragmentation ions was continuously tracked and can be integrated for peak identification. Therefore, the precursor ions can be used for quantification and their corresponding fragmented ions for each peak of precursor ion can be used for identification in combination.

Under the guideline of European SANTE/11813/2017 and Commission Decision 2002/657/EC [22], identification of the concerned analytes with high-resolution mass spectrometry can be performed. The chromatographic information, their mass information should attain given identification points (IPs) to get confirmed results. If the high-resolution mass spectrometric data were collected, 2 IPs are earned if the precursor ion match, and 2.5 IPs for each of their product ions [23,24]. For the identification of all compounds, 4.5 IPs are required. In our work, the *m/z* of isotope, and its relative abundance for precursors were also identified, which leads to higher IPs for structure identification. Therefore, our identification rule should be stricter and more reliable than current regulations, which can result in less false positive result according to our experiment on fortified samples.

2.2. Mobile Phase

Due to the excellent performance of Accucore aQ-MS column in the analysis of multiclass compounds of different polarities, it was employed for chromatography separation of these target compounds. MeOH-water and MeCN-water binary mobile phase were investigated for the separation of the 89 compounds. In order to improve the efficiency of analyte ionization, 5 mM of ammonia formate and 0.1% formic acid (FA) were added in both phases. The result showed no triggered MS/MS spectrum for fenitrothion, chlorpyrifos, phorate, or dichlorvos since the automatic gain control AGC does not satisfy the setting value 5×10^5, when MeCN was applied as mobile phase at the concentration of 50 ng/mL under the full scan/dd-MS2 acquisition, which was considered as a negative result in our experiment. Moreover, signal intensities of more than 10 compounds decreased by 1–2 orders compared with MeOH as the mobile phase. Compounds with significant difference of signal intensity are shown in Figure 1. There were unremarkable differences for the rest pesticides on either mobile phase. According to Figure 1, MeOH is a better mobile phase, as more compounds showed higher response on mass spectrometer. Therefore, MeOH-water system with buffers and formic acid was selected for eluting these compounds from the column, and which is similar to Raina's research concerning of determination OPs in the air based on LC-MS/MS [25]. Neither MeCN nor MeOH could separate 89 pesticides completely. However, with the mass spectrometer, these compounds are not necessarily to be separated, as the different *m/z* can be easily acquired and extracted for different co-eluted compounds, with a pure chromatographic signal for individual compound. It should be noted that proper chromatographic elution of these compounds is still important, as it can avoid matrix effect and potentially competitive ionization between each other if high content compounds are present.

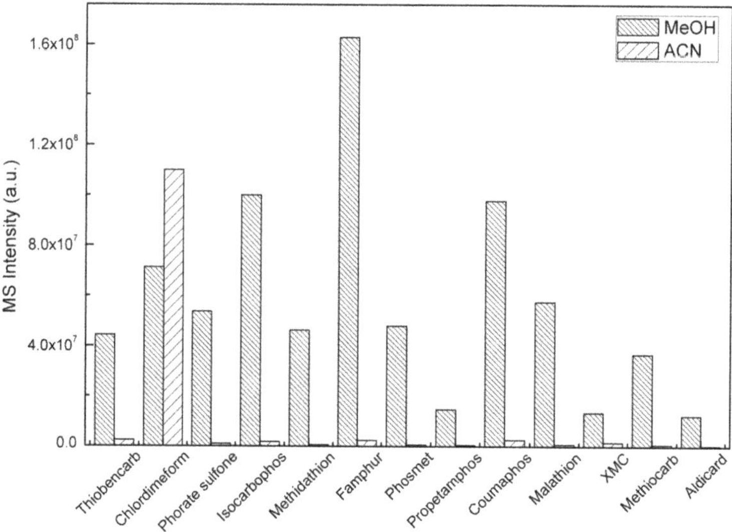

Figure 1. Pesticides with significant changes in sensitivity in MeOH and acetonitrile (ACN) mobile phases with 5 mM ammonium formate and 0.1% formic acid (FA) at the concentration of 50 ng/mL.

2.3. *Buffers*

The addition of formic acid helped improve the ionization efficiency and further increased the sensitivity of analytes, which has been validated in our optimization work. In our research, different concentration of buffers (ammonium formate, 0 mM, 2 mM, and 5 mM) in mobile phase with 0.1% FA were examined for 50 ng/mL mix standards solutions in the same gradient elution. Results showed better chromatographic peaks for most of the compounds when 2 or 5 mM ammonium formate was added in the mobile phase. As it is shown in Figure 2, signals were enhanced by approximately 10 times for propetamphos, famphur, methidathion, and indoxacarb are obtained when buffers were used in the mobile phase. Furthermore, the retention time of some compounds have been delayed after addition of 5 mM ammonium formate in the mobile phase. Buffers are beneficial to the retention and separation for many compounds, especially for acephate, propetamphos, methomyl, and indoxacard, and they further increase the sensitivity, even though a soft/lower intensity on mass spectrum was shown for phorate, dichlorvos, and chlorpyrifos-methyl when 5 mM ammonium formate added. As a result, 5 mM of ammonium formate was added in both mobile phases.

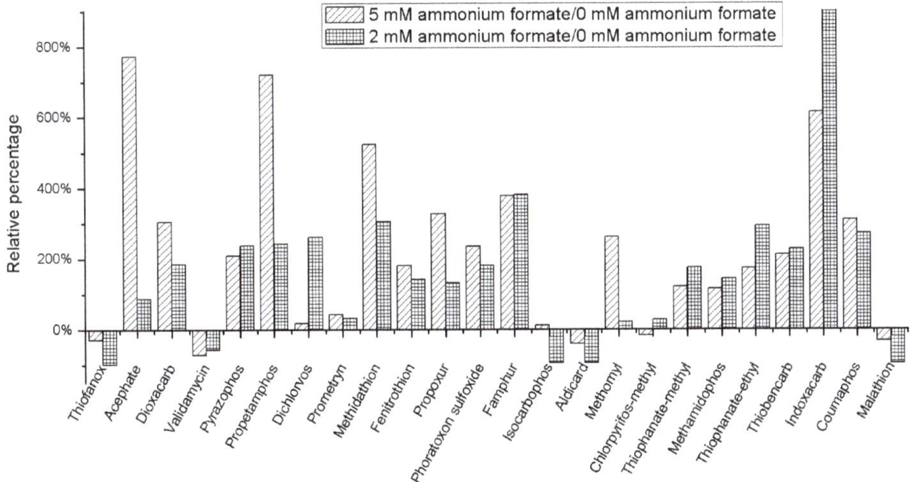

Figure 2. Relative signal enhancement or depression for typical compounds with different concentrations of ammonia formate or without this buffer in the mobile phase in the same gradient.

2.4. Mass Spectrometry

In principle, the higher the resolution of mass spectrum, the identification for target compounds is more accurate. A resolution of 140,000 can be achieved with Orbitrap in our work. However, the analyze time for each scan would be extended significantly and result in a lower data sampling rate. Therefore, enough information for peak integration or critical fragments of precursors will be compromised, as there are only around 15 s of elution time for each compound in chromatography. Similar to our previous work for veterinary drug screening [26,27], full scan/dd-MS2 (TopN) was applied for mass data acquisition, in which an inclusion list of the target compounds was preset. The MS resolution for full scan and fragment acquisition are 70,000 and 17,500, respectively. It could allow the discrimination of low abundant ions undetectable under low resolution [28], and further minimize possibility of false positive [8,29]. For dd-MS2 acquisition, if the high abundant ions were preset in the inclusion list, they were fragmented and scanned sequentially once their precursors were detected. Based on the set parameters, the probability that the instrument fails to trigger MS/MS spectrum acquisition for a detected chemical is greatly reduced. No false negative results were determined for any analyte spiked above its SDL. If there were compounds showing no fragmentation acquisition at the lower concentration, which can be identified by precursor *m/z* abundance greater than 5×10^5, isotope abundance and retention times with narrower deviation to avoid false negatives. Otherwise the compound is counted as undetected. In this work, the top 2 abundant ions were successively fragmented and transferred into the Orbitrap for data acquisition. Under the electrospray ionization, 76 of these compounds formed precursor ions as $[M + H]^+$, 8 of these compounds ionized as $[M + Na]^+$, and 5 pyrethroids formed additions as $[M + NH_4]^+$. PCP Na and 4 phenylpyrazoles formed negative ions as $[M - H]^-$. Three different normalized collision energy (stepped NCE) allowed the high-efficiency fragmentation of different precursors at their best.

2.5. Sample Preparation

It is critical for high recovery determination to choose the solvent of extraction. In this research, pesticides of interest are of multiclass and of quite different chemical or physical properties. To dissolve or extract different analytes with high or low polarity, MeOH and 10% ethyl acetate in MeOH were used as extract solvents for pure Chinese herb drugs, which contains complex matrices and impurities. Results showed better extracting efficiency when MeOH was used. In terms of the recovery of these target compounds, more than half of targets showed better recovery than 10% ethyl acetate in MeOH. As it is shown in Figure 3, seven compounds including phorate, mevinphos, fenobucarb, chlordimeform, propoxur, XMC, and propamocarb showed more than 35% decrease of recovery. Therefore, MeOH was preferred as a solvent for the analysis of pesticides in these drugs.

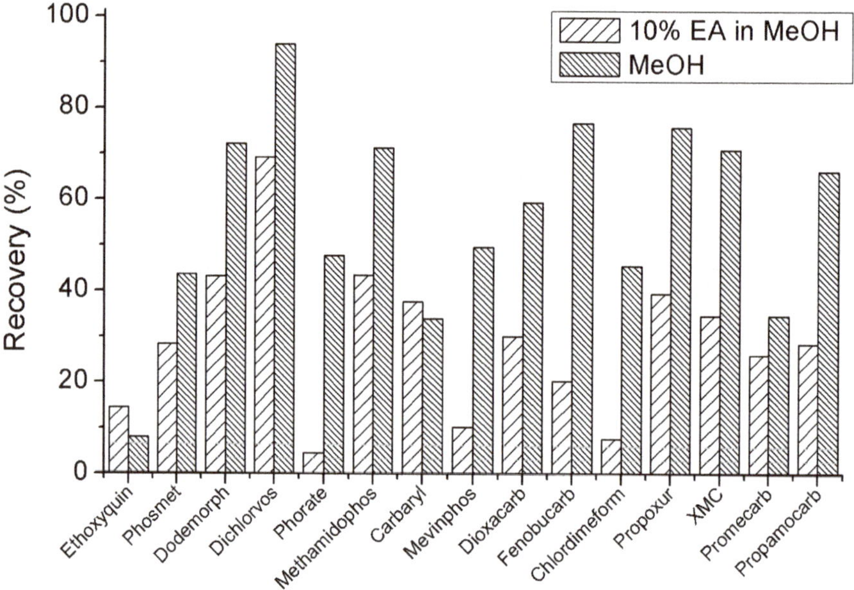

Figure 3. Recoveries of typical compounds with or without the use of 10% ethyl acetate in the methanol.

2.6. Matrix Effect

Matrix effect should be considered in the detection process, which includes intrinsic organic or inorganic compounds after extraction and cleanup, and extrinsic inorganic ions, organic acids, detergent, etc. These interfere material can comprehensively enhance or suppress the response of the target compounds. In our research, the matrix effect (*ME*%) was calculated based on the following equation [30,31]:

$$ME\% = (\frac{A}{B} - 1) \times 100\%$$

where *A* is the integration area in matrix-matched standard solution and *B* is the integration area in a standard solution with identical concentration for each compound. In general, the matrix effect within ±20% can be regarded as acceptable and the calibration can be performed without considering matrix effect. Otherwise, it should be considered during quantification [16,32].

The fishery drugs were dissolved and diluted up to 1000 times, which would significantly decrease the matrix effect. Standards diluted with more than 90% blank matrix solution were used to test for the matrix effect. The result showed less than 20% matrix effect for all the compounds of interest at the concentrations of 100 and 500 ng/mL for compounds with SDL above 100 ng/mL. Because of the acceptable matrix effect, it is feasible to use a methanol–water (1:1, *v*/*v*) solution to dilute a series of standard solutions, for quantification of positive compounds.

2.7. Method Validation

2.7.1. Screening Detection Limit

According to SANTE/11813/2017 [33], the screening detection limit (SDL) was examined with similar process, but less replicates, which has been applied in many reported works [34–36]. Fishery drug of Pure Chinese herb was fortified with mixed standard solutions at different concentrations in six duplicates together with their non-spiked counterparts, which were used for the examination of the screening detection limit, and all compounds satisfied 100% detection criterion at their SDL. Simultaneously, an additional criterion, identity confirmation through the 13C/12C-ratio, was satisfied for each target compound at the corresponding theoretical SDL [34–36]. In our experiment, 1 mg/kg, 10 mg/kg, 50 mg/kg, 100 mg/kg, and 500 mg/kg of these mix target samples were prepared respectively. All these fortified samples were pretreated following the aforementioned method (2.4). Results showed that 54, 80, 85, 86, and 89 compounds were screened positive at 1 mg/kg, 10 mg/kg, 50 mg/kg, 100 mg/kg, and 500 mg/kg, respectively.

2.7.2. Accuracy and Repeatability

The accuracy and repeatability of the screening method were investigated under the fortified concentrations of 10 mg/kg and 100 mg/kg in fishery drug of pure Chinese herb. For compounds at the detection limit of 500 ng/mL on the mass spectrometer, fortified samples of 500 mg/kg were prepared independently. Under the fortified concentration of 10 mg/kg and 100 mg/kg, compounds with the instrument detection limit of 10 ng/mL and below can be readily detected. Over sixty compounds showed the recovery of 70%–110% at spiked 10 and 100mg/kg; fifteen compounds with 110%–120% at 10 mg/kg; twelve compounds with 110%–120% at 100 mg/kg; and three compounds including chlorpyrifos, phosmet, and tributylphos-phorotrithioate had recoveries of over 125% at both spiked levels. Over 95% of compounds identified at both fortified levels had RSD of less than 15%. Compounds were not identified at the lower fortified level but detected at 100 mg/kg including amitraz, phorate, fenitrothion, validamycin, and prothiofos, with the recovery of 59.3%–125% and RSD of 6.17%–14.7%. Compounds only detected at the spiked level of 500 mg/kg are bromophos ethyl, cyfluthrin, parathion, with recovery of 85.3%–105% and RSD of less than 20%. All the quantification results were obtained with less than 20% RSD. Because of the soft matrix enhancement, there were some compounds with high recoveries at both fortified levels for quantification with the standard matched solvent, especially for chlorpyrifos, phosmet, and tributylphos-phorotrithioate. The details of recovery and RSD are presented in Table 1. It is noticed that some compounds did not meet the recovery criteria at one or both of the fortified levels, which could be attributed to high volatility and easy converting properties.

2.7.3. Calibration and linearity

As the matrix effect on the response of the fishery drug sample is quite low, and the recovery results satisfied the semiquantification analysis for most of the compounds in positive samples, the standard solution without matrix matched, and internal standards can be amenable for calibration of positive samples from the perspective of economic costs. In our research, different concentrations of mixed pesticide standards were prepared directly with MeOH–water (1:1, *v*/*v*). Results on mass spectrometer demonstrated that the R-squared of 81 pesticides were no less than 0.990, and 5 other pesticides, including chlorpyrifos, flumethrin, flucythrinate, tau-fluvalinate, and deltamethrin showed R-square between 0.982 and 0.990. The detailed linear profile for 82 compounds is listed in the electronic Supplementary Material (Table S1). The distribution pie chart of the linear range of these compounds is presented in Figure 4.

Table 1. Recovery and relative standard deviation (RSD) of each drug at different spike levels in feedstuff matrices.

Compound	10 mg·kg^{-1}		100 mg·kg^{-1}		Compound	10 mg·kg^{-1}		100 mg·kg^{-1}	
	Recovery (%)	RSD (%, n = 3)	Recovery (%)	RSD (%, n = 3)		Recovery (%)	RSD (%, n = 3)	Recovery (%)	RSD (%, n = 3)
Aminocarb	104	5.99	109	4.35	2,3,5-Trimethacarb	80	1.78	103	0.71
Carbaryl	81.8	3.91	99.9	8.49	3,4,5-Trimethylphenol	86.4	4.62	101	2.91
Carbendazim	120	3.77	85.6	4.04	Acephate	102	4.75	116	4.83
Carbofuran	92.1	6.42	94	3.82	Aldicarb sulfone	94.7	10.2	103	4.07
Chlordimeform	81.7	4.96	85	3.17	Aldicarb sulfoxide	86.6	6.19	103	5.97
Coumaphos	105	7.77	109	8.11	Aldicard	89.9	18.2	105	2.97
Dimethoate	110	11.1	107	10.7	Avermectin B1a	91.4	7.87	84.5	4.24
Dodemorph	95.2	7.08	82.1	2.96	Bendiocarb	80.7	3.5	88.9	7.96
Famphur	113	7.1	111	5.44	Bifenthrin	86.5	21	86.7	5.1
Fenobucarb	82.3	3.33	105	2.89	Chlorpyrifos-methyl	94.5	23	124	10.1
Fuberidazole	111	5.44	85	4.01	Deltamethrin	77.2	9.61	79.2	4.54
Imazalil	92.6	5.67	82.2	4.31	Dichlorvos	113	7.61	117	7.23
Indoxacarb	84.2	7.71	72.9	4.11	Dioxacarb	96	5.6	104	5.34
Isocarbophos	96.9	12.3	104	8.42	Doramectin	75.1	8.76	86.6	4.26
Malathion	88	6.52	113	6.66	Ethoxyquin	94.4	5.48	30.2	16.5
Methiocarb	86.1	8.01	104	5.37	Fenvalerate	90.5	21.6	78.8	1.83
Mevinphos	84.8	10.2	115	1.98	Fipronil	80.8	9.54	89.3	2.1
Monocrotophos	113	5.67	121	4.98	Fipronil-desulfinyl	88	8.67	80.8	6.54
Omethoate	123	12.7	117	12.7	Fipronil-sulfide	83.1	8.68	86.6	3.43
PCP Na	83.1	7.62	88	0.989	Fipronil-sulfone	82.9	6.88	80.8	4.43
Phoratoxon sulfoxide	92.2	14	110	4.59	Flucythrinate	97.9	7.69	80	2.09
Phosalone	105	7.34	103	4.11	Flumethrin	65.3	5.66	77.9	0.732
Phoxim	111	10	102	7.12	Isoprocarb	80	1.78	103	0.71
Pirimicarb	85.5	7.96	99.7	2.84	Ivermectin B1a	59.7	4.64	86.3	1.29
Pirimiphos-methyl	113	8.83	119	3.26	Methamidophos	108	6.42	124	6.08
Promecarb	89.1	5.83	102	2.98	Methidathion	98	4.15	113	4.73
Prometryn	88.9	7.63	78.8	0.618	Methomyl	94	6.7	111	4.89
Propamocarb	96.3	8.1	109	5.37	Phorate sulfone	97.3	5.51	107	1.58
Propazine	93.1	8.22	72.5	2.52	Propetamphos	113	5.06	123	4.27
Propiconazole	115	6.92	77.3	0.198	Robenidine	90	12.2	64.5	1.73

Table 1. *Cont.*

Compound	10 mg·kg^{-1}		100 mg·kg^{-1}		Compound	10 mg·kg^{-1}		100 mg·kg^{-1}	
	Recovery (%)	RSD (%, $n = 3$)	Recovery (%)	RSD (%, $n = 3$)		Recovery (%)	RSD (%, $n = 3$)	Recovery (%)	RSD (%, $n = 3$)
Propoxur	85.2	4.67	91.2	6.94	Tau-fluvalinate	65.5	13.6	83.6	4.04
Pyrazophos	119	12.1	122	6.95	Thiofanox	80.6	2.06	95.4	3.46
Quinalphos	113	8.76	112	5.45	Thiofanox sulphone	87.7	9.49	92.4	0.0845
Simazine	86.7	3.21	65.7	1.7	Thiofanox -sulphoxide	82.9	3.47	101	2.42
Simetryne	98.2	4.56	76	3.28	XMC	75.6	9.88	85.8	6.46
Thiabendazole	119	6.16	80.9	3.43	Xylazine	90.5	8.82	76.8	1.8
Thiobencarb	117	6.87	76.8	1.54	Amitraz	—	—	61.9	14.2
Triazophos	112	5.76	116	3.33	Fenitrothion	—	—	124	13.8
trichlorfon	109	4.85	109	6.27	Phorate	—	—	125	12.1
Chlorpyrifos	138	7.17	154	7.63	Prothiofos	—	—	100	14.7
Phorate sulfoxide	109	9.55	126	3.67	Validamycin	—	—	59.3	6.17
Phosmet	131	10.5	177	17	Bromophos ethyl *	—	—	85.3	19.4
Thiophanate-ethyl	128	5.49	76.9	1.36	Cyfluthrin *	—	—	91.5	1.13
Thiophanate-methyl	128	4.96	77.9	1.86	Parathion *	—	—	105	3.53
Tributylphos-phorotrithioate	150	7.98	147	4.56					

* fortified at 500 mg/kg.

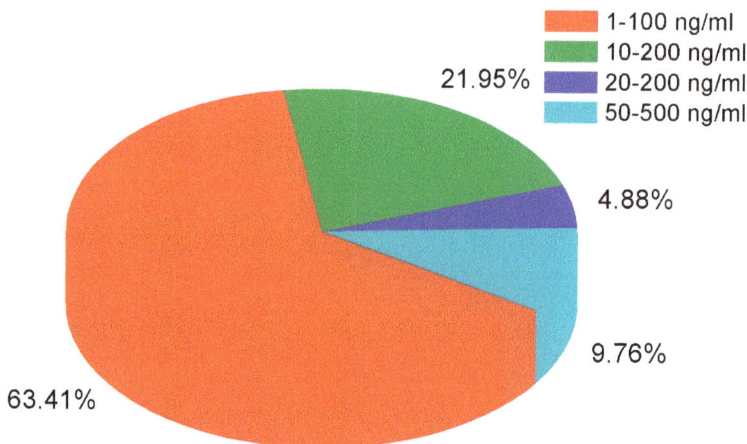

Figure 4. The percentages of different linear range of 82 targeted pesticides.

2.8. Practical Screening

The method was further applied in screening of 21 fishery drug samples (pesticides, water-clean agents and antibacterial agents). Samples were prepared according to sample preparation (4.4) prior to analysis. For the compounds with concentration out of the linear range, samples were re-diluted with the dilution factor of samples adjusted to ensure the concentration to be quantitatively evaluated based on our linear range. The screening was carried out following the home-built database and preset rules. Quantification was conducted through the peak areas of precursor ions in positive samples and was externally calibrated. Based on the database and preset identification rules, 10 out of the 21 fishery drug samples were screened positive with pesticides. 10 samples were detected with unspecified components. As is shown in Table 2, the identified pesticides were chlorpyrifos, ivermectin B1a, phoxim, avermectin B1a, and carbendazim. Three samples contained forbidden drug chlorpyrifos (Figure 5A), and 5 samples contain avermectin and ivermectin (Figure 5B) of more than 3 g/L. Their chromatographic and fragment information was highly identical to the standards, as are shown in Figure 5. Detailed information of the screened positive samples was presented in Table 2.

Table 2. Screening results of practical fishery drugs.

Code	Trade Name	Dosage Form	Listed	Detected Compounds	Contents (mg/kg or mg/L)
4	Insecticide for fish	Aqueous solution	NA	Chlorpyrifos	2.66
5	Insecticide for fish and shrimp	Soluble concentrate	Avermectin	Ivermectin B1a Chlorpyrifos Avermectin B1a	347 1.33 7479
6	Pesticide for water	Soluble concentrate	Bioactive ingredient	Ivermectin B1a Avermectin B1a	207 3482
14	Insecticide for fish	Soluble concentrate	NA	Phoxim	2.20
15	Avermectin solution	Soluble concentrate	Avermectin	Avermectin B1a	5937
16	Benzalkonium Bromide Solution	Aqueous solution	NA	Avermectin B1a	55,587

Table 2. *Cont.*

Code	Trade Name	Dosage Form	Listed	Detected Compounds	Contents (mg/kg or mg/L)
17	Beta-Cypermethrin Solution	Aqueous solution	Cypermethrin	Chlorpyrifos	9.11
20	Insecticide for water	Soluble concentrate	Bioactive ingredient	Ivermectin B1a	8214
21	Pesticide for water	Gel solution	Avermectin	Ivermectin B1a Avermectin B1a	121 3736
22	Insecticide for water	Soluble concentrate	Avermectin	Phoxim Avermectin B1a	19.7 1931

NA: not available. Listed: active compounds were listed in the label of fishery drugs.

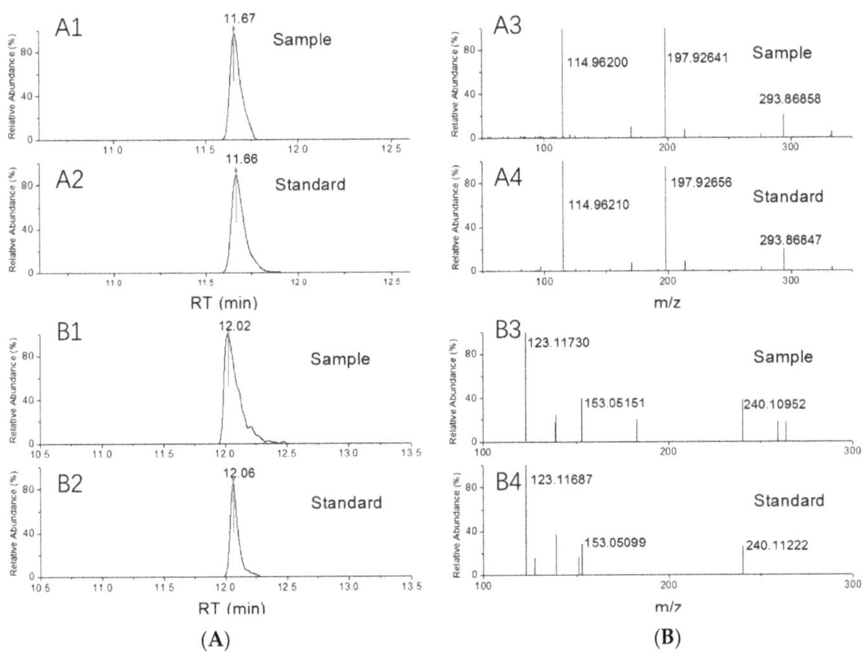

Figure 5. Comparison of the spectra of chlorpyrifos (**A**) and ivermectin B1a (**B**) detected in the positive samples. A1, B1 and A2, B2 are the chromatogram of real samples and standards, respectively. A3, B3 and A4, B4 is the MS/MS spectrum of the positive sample and the standard respectively.

3. Materials and Methods

3.1. Instruments and Reagents

The ultrahigh-performance liquid chromatography (UHPLC) system (Dionex UltiMate 3000, Thermo Fisher Scientific, San-Francisco, USA) coupled to quadrupole Orbitrap mass spectrometer with electrospray ionization (Q-Exactive, Thermo Fisher Scientific) was used for data acquisition.

Eight-nine Pesticides were selected for target screening as listed in Table 3. Carbofuran and dichlorvos were obtained from MANHAGE Biotech. Inc. (Beijing, China), thiofanox-sulfone, thiometon, aldicarb-sulfone, phoratoxon sulfoxide, PCP Na were purchased from Accustandard Inc. (New Haven, CT, USA) The other 85 pesticides standards were supplied by Dr. Ehrenstorfer GmbH (Augsburg, Germany). Acetonitrile (MeCN) and Methanol (MeOH) of HPLC grade were obtained from J.T. Baker

(Phillipsburg, NJ, USA). Formic acid (FA, 98%, LC-MS grade, Fisher Scientific, Spain, or HPLC grade) was obtained from FLUKA. All the other relevant reagents were purchased from common domestic suppliers. Pure water was obtained through Water Milli Q ELEMENT purification unit (Millipore, Bedford, USA).

3.2. Preparation of Standards

Standards stock solution: c.a. 5 mg solid standards was dissolved with MeOH in 10 mL beaker, and then transferred to a 50 mL flask and diluted with MeOH. For compounds will less solubility in MeOH, 0.1 mL formic acid (98%, HPLC grade) was firstly added and the mixture was sonicated until the solids were completely dissolved. Five microliters of liquid standards were pipetted into a 10 mL-beaker and weighed to get accurate mass. After that, they were dissolved in methanol following similarly procedure as the solid standards. All these single standards stock solutions were c.a. 100 μg/mL. The purchased standards solutions were not diluted until further preparation of mixed standards solution. Mixed standards solutions were prepared by mixing standards of the same category, which were finally diluted to 5 μg/mL. The standards were categorized into organophosphorus, carbamate, organochlorine, imidazole, pyrethroid, triazole, phenylpyrazole, avermectin, and miscellaneous. All the standards solutions were stored in refrigerator at −42 °C.

Matrix-matched standards were used to evaluate the matrix effect, where the standards were dissolved into a matrix of Chinese herbal fishery drug, which was negative for pesticides before spiking.

3.3. Methods

3.3.1. Elution Conditions

Accucore aQ-MS column (2.1 × 100 mm, 2.6 μm, Thermo Fisher Scientific, USA), was employed to perform sample separation with a thermostat at 30 °C. The binary mobile phases (MP) were 0.1% FA in water (containing 5 mM Ammonium Formate, A) and 0.1% FA in MeOH (containing 5 mM Ammonium Formate, B). Their gradient elution was started with 2% B, linearly increasing to 20% in 4 min and continuously ramped to 40% within 1.5 min. Subsequentially, B was increased to 98% in the subsequent 5 min, and kept for 2.4 min. Then, B was restored to the initial conditions 2%B in the following 2.1 min, and kept for 5 min to re-equilibrate for the next injection. The whole elution process for one injection analysis took 20 min. The flow rate was kept at 0.3 mL·min^{-1}. The injection volume for analysis was 10 μL for each sample.

3.3.2. Mass Spectrometer Condition

Parameters for electrospray were as following: spray voltage, 3200 V (positive mode), 2800 V (negative mode), sheath gas flow rate at 40 L·min^{-1}, auxiliary gas flow rate at 10 L·min^{-1}, sweep gas flow rate at 1.0 L·min^{-1}, auxiliary gas temperature at 350 °C, capillary temperature at 325 °C and S-lens RF level at 60 V. The scan mode for high-resolution mass spectrometry acquisition was Full MS/dd-MS2 (with inclusion list) mode. Recorded mass range for full mass record was between m/z 100–1000 (positive mode) and 150–1000 (negative mode), at resolution of 70,000. The Full MS/dd-MS2 (with inclusion list) mode can simultaneously record the precursor mass and the MS/MS (fragmentation) spectra for selected precursors. The MS/MS acquisition for fragment scanning of the selected ions was carried out at the isolation window of 2.0 m/z and the resolution of 15,000. For each round of fragmentation acquisition, the top 2 (TopN, 2, loop count 1) abundant precursors above the threshold 5×10^5 were sequentially transferred into the C-Trap (AGC, 5×10^5, Max IT, 100 ms) for collision at normalized energies (NCE, 20, 50, 80) in HCD multipole and pumped to Orbitrap for MS/MS acquisition.

All the units of UHPLC-ESI-Q-Orbitrap HRMS system were controlled through the Tracefinder software.

Table 3. Chromatographic, mass spectrometric information and limit of detection of 89 targeted pesticides.

Compounds	RT (min)	Extracted Mass (*m*/*z*)	Fragment Ions (*m*/*z*)	LOD ng/mL	Compounds	RT (min)	Extracted Mass (*m*/*z*)	Fragment Ions (*m*/*z*)	LOD ng/mL
Bendiocarb	8.95	224.09173	167.07027/109.02841	1	Aldicarb sulfone	5.15	223.0747	86.06004/76.03930	1
Chlorpyrifos	11.75	349.93356	114.9615/197.92730	1	Aminocarb	4.39	209.12845	152.10699/137.08352	1
Coumaphos	10.96	363.02174	226.99263/306.95913	1	Carbaryl	9.27	202.08626	132.04439/124.08827	1
Dimethoate	7.44	230.0069	142.99623/170.96978	1	Carbendazim	6.3	192.07675	160.05054/132.05562	1
Dodemorph	9.78	282.27914	116.10699/98.09643	1	Carbofuran	8.94	222.11247	123.04406/165.09101	1
Famphur	9.55	326.02803	93.00999/142.99263	1	Chlordimeform	6.63	197.084	117.05730/152.02615	1
Fuberidazole	7.4	185.07094	157.07602/156.06820	1	Dioxacarb	7.46	224.09173	123.04406/167.07027	1
Isocarbophos	9.68	312.04299	269.99844/236.00554	1	Ethoxyquin	9.7	218.15394	148.07569/190.12264	1
Malathion	10.26	353.02529	195.06031/227.03238	1	Fenobucarb	10.02	208.13321	208.13321/95.04914	1
Mevinphos	8.12	225.05225	127.01547/67.01784	1	Fipronil	10.62	434.93143	329.95845/183.01646	1
Monocrotophos	6.65	224.06824	127.01547/58.02874	1	Fipronil-desulfinyl	10.54	386.96444	350.98667/281.99146	1
Omethoate	4.17	214.02974	214.02974/182.98754	1	Fipronil-sulfide	10.72	418.93651	57.9746/170.00863	1
Phorate sulfoxide	9.47	277.01502	114.96133/142.93848	1	Fipronil-sulfone	10.85	450.92634	414.94857/243.98839	1
Phoratoxon sulfoxide	7.61	261.03786	114.96133/128.97698	1	Imazalil	9.49	297.0556	158.97628/69.04472	1
Phosalone	11.06	367.99414	114.96150/138.01033	1	Indoxacarb	11.14	528.07799	203.01866/168.02107	1
Phosmet	9.98	318.00181	160.03930/133.02841	1	Methiocarb	10.14	226.08963	169.06816/121.06479	1
Phoxim	11.1	299.06138	129.04472/95.04945	1	Pentachlorophenol	11.74	262.83973	262.83973/262.83973	1
Pirimiphos-methyl	11.02	306.10358	108.05562/67.02907	1	Pirimicarb	8.38	239.15025	182.12879/72.04439	1
Prometryn	10.22	242.14339	200.09644/158.04949	1	Promecarb	10.29	208.13321	208.13321/109.06479	1
Propazine	10.1	230.1167	188.06975/146.02280	1	Propamocarb	4.25	189.15975	102.05496/74.02365	1
Pyrazophos	11.08	374.0934	194.05602/222.08732	1	Propiconazole	10.96	342.07706	158.97628/69.06988	1
Quinalphos	10.84	299.06138	147.05529/163.03245	1	Propoxur	8.89	210.11247	111.04406/95.04914	1
Simazine	9.01	202.0854	132.03230/124.08692	1	Robenidine	10.38	334.06208	138.01050/155.03705	1
Simetryne	9.13	214.11209	96.05562/124.08692	1	Thiobencarb	11.14	258.07139	258.07139/125.01525	1
Thiabendazole	7.26	202.04334	175.03245/131.06037	1	Thiophanate-ethyl	9.68	371.08422	151.03245/282.03654	1
Triazophos	10.42	314.07228	162.06619/114.96133	1	Thiophanate-methyl	8.85	343.05292	151.03245/311.02671	1
Tributyl-phosphorotrithioate	12.11	315.10344	57.06988/168.99052	1	Xylazine	7.26	221.1107	90.03720/164.05285	1
Trichlorfon	7.2	256.92985	127.01547/220.95318	1	2,3,5-Trimethacarb	9.56	194.11756	135.04406/107.04914	2
Doramectin	12.29	921.49708	777.41843/449.25097	2	3,4,5-Trimethylphenol	9.75	194.11756	135.04406/107.04914	2
Isoprocarb	9.56	194.11756	137.09609/95.04914	2	Phorate sulfone	9.55	293.00993	246.96807/114.96133	2
Methidathion	9.86	302.96913	145.00662/71.02399	2	Thiofanox sulphoxide	7.14	235.11109	57.06988/104.01646	2
XMC	9.39	180.10191	123.08044/113.99744	2	Avermectin B1a	12.23	895.48143	123.11683/153.05462	5
Aldicard	8.32	213.06682	95.04914/141.00593	5	Deltamethrin	12.04	521.00699	278.90762/89.05971	5
Dichlorvos	7.18	220.95318	127.01565/78.99452	5	Flucythrinate	11.73	469.19334	114.09134/199.0929	5
Propetamphos	10.34	282.09234	138.01370/156.02395	5	Flumethrin	12.33	532.0853	114.09134/73.02982	5

Table 3. *Cont.*

Compounds	RT (min)	Extracted Mass (m/z)	Fragment Ions (m/z)	LOD ng/mL	Compounds	RT (min)	Extracted Mass (m/z)	Fragment Ions (m/z)	LOD ng/mL
Thiofanox	9.56	241.09812	58.06513/184.07906	5	Ivermectin B1a	12.48	897.49708	753.41843/329.20872	5
Thiofanox sulphone	7.3	251.106	251.10600/57.06988	5	Methamidophos	2.57	142.00861	142.00861/112.01598	5
Bifenthrin	12.45	440.15987	181.10118/166.07770	10	Amitraz	11.97	294.19647	163.12298/122.09643	20
Phlorpyrifos-methyl	11.29	321.90226	142.99263/289.87605	10	Tau-fluvalinate	12.13	503.13438	181.06479/114.09134	20
Acephate	3.46	184.01918	113.00250/142.99263	20	Fenitrothion	10.18	278.02466	142.99263/149.02332	50
Aldicarb sulfoxide	4.78	207.07979	89.04195/69.05730	20	Methomyl	5.84	163.05357	88.02155/106.03211	50
Fenvalerate	12.05	437.16265	437.16265/114.09134	20	Phorate	11.11	261.0201	261.02010/75.02630	50
Validamycin	1.26	498.21812	142.08626/124.07569	50	Prothiofos	12.22	344.97009	258.92025/132.96046	100
Bromophos ethyl	12.21	392.8878	336.82520/161.96337	500	Cyfluthrin	12.01	451.0986	191.00250/114.09134	500
Parathion	10.86	292.04031	114.96133/138.00081	500					

3.4. Sample Preparation

A 20 mL centrifuge tube was filled with 100 mg samples and added with 20 mL MeOH. One hundred microliters of liquid sample was pipetted directly into another centrifuge tube. The tube was vortexed for 1 min, and ultrasonicated for 15 min. Then, the solution was vortexed again and silenced for 2 min. Then, MeOH-water (1:1, *v*/*v*) was used to dilute 0.5 mL of supernatant by 5 fold. After vortex and silence, 1 mL of the solution was transferred to an Eppendorf tube (1.5 mL) and centrifuged at 10,000 rpm/min for 15 min to remove the precipitate. The upper supernatant was transferred into a vial for analysis.

Chinese herbal drugs were used for method validation in the research, which was representative of a complex matrix of fishery products. Fishery drugs of pure Chinese herb products composed of granular herbal extract were purchased from a local fishery store.

3.5. Database for Screening, Qualitative and Quantitative Rules

The names, categories, CAS numbers, formulas, expected mass of the suspected compounds were searched and collected to establish a basic database. Then, the standard solution of each compound (100 ng·mL^{-1}) were analyzed through the UHPLC-ESI-Q-Orbitrap HRMS system using the aforementioned parameters. Therefore, *m*/*z* of precursor ion, retention time (RT) and fragment ions (FI) were acquired by experiments. In parallel, the isotope pattern for each precursor was automatically calculated by Tracefinder software. All information was organized and built in Tracefinder. It was used to perform screening according to the database with the following screening rules: *m*/*z* deviation of precursor ion was 3×10^{-6}, allowed RT deviation was ±15 s, at least one fragment ion match with allowed *m*/*z* deviation at 2×10^{-5}, and the fit threshold for precursor isotope pattern was more than 75% with allowed mass deviation within 10 ppm, and allowed isotope intensity deviation of less than 25%. If the screening rules passed for a compound, it was qualitatively identified as positive. Furthermore, a series of mixed standards solution of 1–500 ng/mL were prepared for quantification of the positive compounds. The integrated peak area of precursor ions for positive compounds was used for external calibration and quantification. The instrument detection limit (LOD), the minimum concentration that the compound could be identified under the qualitative rules, was tested at the optimized parameters. The detailed information for these compounds of interest in the database is shown in Table 3.

4. Conclusions

In summary, a database of 89 pesticides was built, including both chromatographic and HRMS information. The data was acquired after parameter optimization on ultrahigh performance liquid chromatography interfaced quadrupole-Orbitrap mass spectrometer. Based on the database, the screening rule for these compounds was further established by comparing their precursors, fragments, retention time and isotopes. The fast, high-throughput identification and rough quantification of these compounds was achieved. The method was successfully applied for the pesticides risk assessment of fishery drugs. However, as the detection mode established on potential and known pesticides, where their chromatographic and mass spectrometric information were examined and collected, the unknown, non-target risk compounds were ignored. Further work will be focused on non-target screening based on characteristic fragments for recognizing and monitoring risk factors. Overall, our current method can be used as a fast, reliable, efficient and practical tools for the fishery drug risk assessment, which saves more time, and expenses.

Supplementary Materials: The following is available online, Table S1: The detailed linear profile for 82 compounds.

Author Contributions: C.K. and H.-J.Y. conceived and designed the experiments; S.-Y.W. and Q.-P.C. performed the experiments; C.K. and S.-Y.W. analyzed the data; C.K. and S.-Y.W. wrote the paper.

Funding: This research was funded by the National Natural Science Foundation of China (31701698), Shanghai Agriculture Applied Technology Development Program (Grant No. T2016010401), Shanghai Key Laboratory of Forensic Medicine (Academy of Forensic Science, KF1910).

35. De Paepe, E.; Wauters, J.; Van Der Borght, M.; Claes, J.; Huysman, S.; Croubels, S.; Vanhaecke, L. Ultra-high-performance liquid chromatography coupled to quadrupole orbitrap high-resolution mass spectrometry for multi-residue screening of pesticides, (veterinary) drugs and mycotoxins in edible insects. *Food Chem.* **2019**, *293*, 187–196. [CrossRef]
36. Huysman, S.; Van Meulebroek, L.; Vanryckeghem, F.; Van Langenhove, H.; Demeestere, K.; Vanhaecke, L. Development and validation of an ultra-high performance liquid chromatographic high resolution q-orbitrap mass spectrometric method for the simultaneous determination of steroidal endocrine disrupting compounds in aquatic matrices. *Anal. Chim. Acta* **2017**, *984*, 140–150. [CrossRef]

Sample Availability: Samples of the compounds are not available from the authors.

Review

Determination of Kanamycin by High Performance Liquid Chromatography

Xingping Zhang [1,2], Jiujun Wang [1], Qinghua Wu [1], Li Li [1], Yun Wang [1] and Hualin Yang [1,2,*]

[1] College of Life Science, Yangtze University, Jingzhou 434025, China; xpzhang1987@163.com (X.Z.); wangjiujun2018@126.com (J.W.); wqh212@hotmail.com (Q.W.); lily2012@yangtzeu.edu.cn (L.L.); 1wangyun@yangtzeu.edu.cn (Y.W.)

[2] Research and Development Sharing Platform of Hubei Province for Freshwater Product Quality and Safety, Yangtze University, Jingzhou 434025, China

* Correspondence: yanghualin2005@126.com; Tel./Fax: +86-716-8359-2572

Academic Editors: Marcello Locatelli, Angela Tartaglia, Dora Melucci, Abuzar Kabir, Halil Ibrahim Ulusoy and Victoria Samanidou

Received: 1 April 2019; Accepted: 15 May 2019; Published: 17 May 2019

Abstract: Kanamycin is an aminoglycoside antibiotic widely used in treating animal diseases caused by Gram-negative and Gram-positive infections. Kanamycin has a relatively narrow therapeutic index, and can accumulate in the human body through the food chain. The abuse of kanamycin can have serious side-effects. Therefore, it was necessary to develop a sensitive and selective analysis method to detect kanamycin residue in food to ensure public health. There are many analytical methods to determine kanamycin concentration, among which high performance liquid chromatography (HPLC) is a common and practical tool. This paper presents a review of the application of HPLC analysis of kanamycin in different sample matrices. The different detectors coupled with HPLC, including Ultraviolet (UV)/Fluorescence, Evaporative Light Scattering Detector (ELSD)/Pulsed Electrochemical Detection (PED), and Mass Spectrometry, are discussed. Meanwhile, the strengths and weaknesses of each method are compared. The pre-treatment methods of food samples, including protein precipitation, liquid-liquid extraction (LLE), and solid-phase extraction (SPE) are also summarized in this paper.

Keywords: Kanamycin; HPLC; sample pre-treatment; different detectors; food contamination

---

1. Introduction

Kanamycin is widely used in the treatment of animal infections, added as growth promoters or feed additives for preventive therapy [1]. The antibacterial mechanism of kanamycin is that it can irreversibly bind to the bacterial ribosomal 30S subunit and inhibit its protein synthesis [2]. Because of its potential ototoxicity and nephrotoxicity [3–6], the indiscriminate use of kanamycin will enhance bacterial resistance and cause kanamycin-residue accumulation in animal-derived food, which threatens human health. Therefore, the European Union has promulgated regulations on the maximum residue limits (MRLs) of kanamycin in different food matrices (100 μg/kg for muscle, 100 μg/kg for egg, 600 μg/kg for liver, 2500 μg/kg for kidney, 150 μg/kg for milk and 100 μg/kg for chicken meat) [7].

Kanamycin was isolated in 1957 [8]. It is a mixture of several closely related compounds, such as main constituent kanamycin A (>95%), as well as minor constituents kanamycin B, C, and D (<5%). The major components are shown in Figure 1 [9]. In addition, degradation products such as 2-deoxystreptamine and paromamine can also be present [10]. Kanamycin A and C are isomers, whereas kanamycin B and D have different functional groups [9].

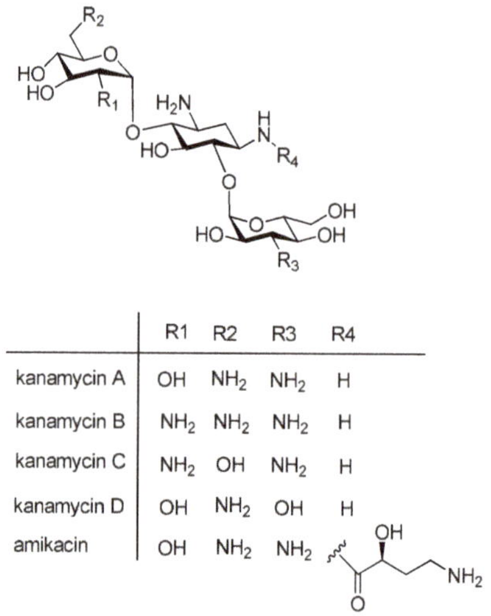

	R1	R2	R3	R4
kanamycin A	OH	NH_2	NH_2	H
kanamycin B	NH_2	NH_2	NH_2	H
kanamycin C	NH_2	OH	NH_2	H
kanamycin D	OH	NH_2	OH	H
amikacin	OH	NH_2	NH_2	

Figure 1. Structure of kanamycin A, B, C, and D and amikacin.

2. The Pre-Treatment Methods of Food Sample

The key point of detecting kanamycin is to remove the impurities or extract kanamycin from matrices. The usual techniques for extraction and cleanup of kanamycin from matrices include protein precipitation, liquid-liquid extraction (LLE), and solid-phase extraction (SPE) [11]. Based on these techniques, pre-treatment methods for kanamycin detection in food samples are summarized as follows.

2.1. Protein Precipitation

Deproteinization was commonly used in the extraction of kanamycin from biological matrices because removal of interferences is necessary to retain good recoveries. Acetonitrile, acidified methanol, and trichloroacetic acid were commonly used precipitation reagents.

In human plasma sample, the simple organic solvent of acetonitrile was used for deproteinization with kanamycin recovery range from 92.3% to 100.8% [12,13]. The acidified methanol with a final concentration of 0.13 mol/L hydrochloric acid (HCl) can also be used for deproteinization of human plasma, and kanamycin recovery ranges from 91.2% to 93.4% [14].

In rat plasma samples, trichloroacetic acid (TCA) with a final concentration of 25–30% was a good precipitation reagent and offers best recovery [15].

In human serum sample, the acidified methonal with a final concentration of 0.14 mol/L HCl can be used to extract kanamycin [16]. Meanwhile, TCA with a final concentration of 40% can be applied in human serum deproteinization, and recovery of kanamycin ranges from 93.9% to 98.4% [17].

Dried blood spots (DBSs) were more convenient than traditional venous blood sampling. In one anti-TB drug analysis, 0.1 mol/L HCl in mixed methonal solution was used for deproteinization of DBS samples [18].

Pig feeds samples were extracted with 0.1 mol/L HCl and kanamycin recovery ranged from 89.4% to 92.8% [19].

In bovine milk, swine and poultry muscle, samples were first precipitated by 15% TCA and then purified with bulk C18 resin. The recoveries of kanamycin were 92% in milk and 36.8–67% in muscle [20].

The chicken meat samples were extracted and precipitated with a mixture of acetonitrile (ACN)-2% TCA (45:55, *v*/*v*), followed by on-line clean-up using turbulent flow chromatography [21]. This automated on-line technique enabled a larger number of samples to be analyzed per day than the traditional clean-up technique. Kanamycin recovery ranged from 109% to 120% in chicken meat [21].

2.2. Liquid-Liquid Extraction

Liquid-liquid extraction (LLE) has been exploited as an extraction procedure for kanamycin from complex matrices. In a published method, veal muscle samples were extracted using CH_3CN-H_2O (86:14 *v/v*), followed by a defatting step using hexane liquid-liquid extraction [22].

A new pre-preparation technique of dispersive liquid-liquid microextraction based on solidification of floating organic droplet (DLLME-SFO) was developed, which is a new kind of LLE method that could be applied in the analysis of volatile and polar compounds, like kanamycin. In wastewater and soil, kanamycin is extracted with dodecanol (extraction solvent) and ethanol (dispersive solvent) [23]. Compared with conventional sample preparation methods, the proposed derivatization followed by DLLME-SFO procedure significantly reduced the consumption of organic solvent with high enrichment factor. The DLLME-SFO method facilitated high extraction efficiency and further wide linear range, with good precision, and lower detection limit. The recovery was found to be between 91.3–102.7% for wastewater and 90.3–107.7% for soil. The linearity range was 0.5–500 ng/mL. The LOD was 0.012 ng/mL and LOQ was 0.05 ng/mL [23].

2.3. Solid-Phase Extraction

In many cases, solid-phase extraction (SPE) have been extensively used to extract and concentrate trace organic materials from samples [24–26]. According to packing materials, the solid phase extraction can be classified into four types: Bonded silica gel particle, high polymer material, adsorptive packing material, and mix-mode and specialized column. In this review, the sorbents used for kanamycin analysis mostly belong to the bonded silica gel type, except for molecularly imprinted polymers (MIPs) [27,28] and Chromabond HR-XC [29], which are a high polymer type sorbent.

According to different retention mechanisms, the SPE sorbents used in this review could be further classified into reversed phases sorbents, ion exchange phases sorbents (cation exchanger and anion exchanger), and normal phases sorbents, as shown in Figure 2. The SPE sorbents included in this review are reversed phase sorbents (ODS-C18, Sep-pak tC18, Oasis HLB), strong cation exchanger (Oasis MCX, Chromabond HR-XC), and weak cation exchanger (WCX, CBA, CBX).

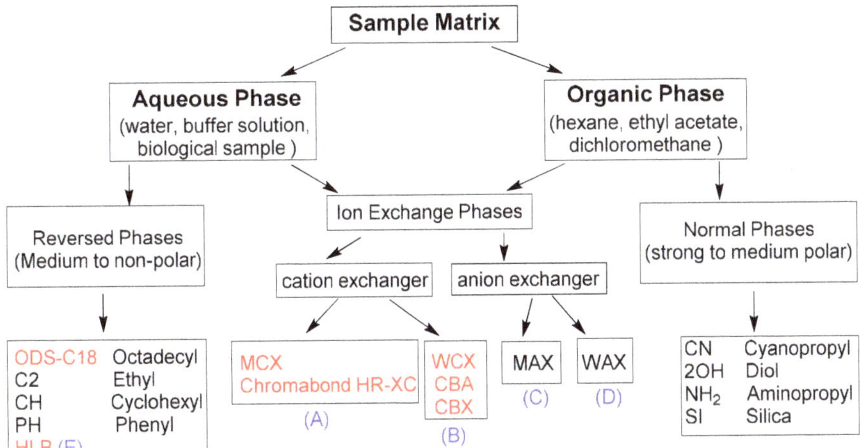

Figure 2. The classification and choice of solid-phase sorbents. (A) Strong cation exchanger; (B) Weak cation exchanger; (C) Strong anion exchanger; (D) Weak anion exchanger; (E) Hydrophilic-lipophilic balanced co-polymer-reversed phased retention.

The reversed phase sorbent Sep-pak tC18 [30] and ODS-C18 [31] was able to extract the non-polar compound from the aqueous sample. The porous silica particles surface bonded with C18 or other hydrophobic alkyl groups. Because of its hydrophobic distribution mechanism, it has strong retention with hydrophobic compounds, but weak retention with hydrophilic compounds. Before use, the cartridge must be preconditioned with a water-soluble organic solvent to solvate the alkyl chains, and then equilibrated with water. It must then be loaded with aqueous samples, followed by eluting with water. A drawback is that before loading the sample, the sorbent must be kept wet, otherwise it will result in low analyte recovery or poor reproducibility. The AccuBOND ODS-C18 cartridge was used for cleanup in soil samples with a kanamycin recovery range from 72.3% to 92.5% [31].

The HLB cartridge has both hydrophilic and lipophilic functional groups, which is a new hydrophilic-lipophilic balanced wettable reversed-phase sorbent [32]. It can overcome the limitations of traditional reversed phase sorbents. Firstly, it is wettable with water, so it has good recovery and reproducibility even the cartridge runs dry during processing. Secondly, it is available for a wide range of compounds including both polar and non-polar chemicals. In muscle, kidney, liver, honey and milk samples, kanamycin was extracted through two consecutive Oasis HLB cartridges (3 mL/60 mg) with a recovery range from 71% to 104% [33].

Ion exchange sorbents (MCX, WCX, MAX, WAX) were found to extract ionizable compounds from the aqueous sample. Because of the ion exchange and hydrophobic distribution mechanisms, the ion exchange sorbents have a strong retention to ionic compounds that have the opposite electric charge of the sorbent carrier, but very weak retention to other compounds [34].

The MCX cartridge is a mixed-mode reversed-phase strong cation exchanger with a pKa of less than 1.0; its sulfonic acid groups have high selectivity to alkaline compounds. Prior to use, it was preconditioned with MeOH, followed by water, then loaded with the extracted sample. Kanamycin is a weak alkaline compound with a pKa of 7.2. At pH lower than 5, the kanamycin was essentially charged and absorbed in the cation cartridge; thus, the sample was extracted with strong acid of 0.1 mol/L HCl [35], 10% TCA [36] or 9% FA [37] aqueous solution prior to loading. At pH higher than 9.0, the kanamycin was neutralised, and the elution could take place. Thus, ammonium methanol solution (1–25%, pH 9.2) was applied to elute kanamycin from the sorbent.

The MCX cartridge was used to extract samples in animal feeds [35], swine tissue [36] and human serum [37] with a kanamycin recovery of 98.4–106% [35] and 80.7% to 91.3% [36], respectively.

The Chromabond HR-X cartridge was styrene-divinylbenzene copolymer based strong cation exchanger. Its surface bonded to benzenesulfonic acid groups [38]. Thus, its retention mechanism was similar to the Oasis MCX sorbent. It was used for cleanup in muscle, kidney and milk samples, with kanamycin recovery ranging from 95% to 102% [29].

The WCX cartridge is a mixed-mode reversed-phase weak cation exchanger with pKa of about 5.0. Its carboxyl groups have high selectivity to strong alkaline compounds. Prior to use, it was preconditioned with MeOH, followed by water, then loaded with the extracted sample. At pH over 6.5, this sorbent was essentially charged to retain kanamycin, so the PH of extracted sample was adjusted to 6.5~7.5 with NaOH and HCl prior to loading. At pH lower than 3.0, the charge on the sorbent was neutralised, and the elution could take place. So, ammonium formate buffer solution (pH 3) [39–41] orformic acid 10% [42], 40% methanol solution [30] were applied to elute kanamycin from the sorbent.

The WCX cartridge (Accell plus CM) was used for cleanup in honey and kidney samples, with kanamycin recovery range from 82% to 105% [40]. The Taurus WCX cartridge was used in honey, milk and liver samples, with a kanamycin recovery range from 58% to 96% [41]. Consecutive SPE cleanup using Sep-pak tC18 and Oasis WCX were applied in milk sample, with a reduced matrix effect and improved absolute kanamycin recoveries from 69.9% to 77.9% [30]. Lehotay et al. used DPX SPE (conducted in a pipet tip rather than a cartridge or centrifuge tube) with 5 mL tips (10 per row) containing 50 mg WCX sorbent for the cleanup of bovine kidney, liver, and muscle samples. The recovery of kanamycin was from 82% to 94% at a spiking level of 0.1 μg/g [42].

The carboxylic acid (CBA) cartridge was a weak cation exchanger with pKa of about 4.8, similar to the Oasis WCX cartridge. Ammonium acetate (pH 7.0) was chosen as the conditioning solution.

The pH of the extracts was adjusted to pH 7.5. The 2% FA in methanol was applied to elute kanamycin from the sorbent. It was used to purify the extracts in muscle, liver, kidney, milk and egg samples [43].

The carboxypropyl (CBX) cartridge was a weak cation exchanger, similar to the Oasis WCX cartridge. The pH of the tissue extract was adjusted to pH 7.0, and then passed slowly through the CBX column. The column was washed with water and then eluted with 5 mL of acetic acid-water-methanolmixture (1:1:8) to get kanamycin; final recoveries range from 81.1% to 104% [44].

Recently, novel sorbents such as molecularly imprinted polymers (MIPs) have emerged [28,45,46]; they are synthetic materials that provide complementary binding sites to specifically capture the target analyte kanamycin. Thus, they are ideal for selective extraction and to reduce the matrix effect. MISPE-Aminoglycoside cartridges (50 mg, 3 mL) were used for extraction and clean-up processes for honey, pork and milk samples, achieving kanamycin recoveries within 78.2–97% and 70–106%, respectively [27,28]. The matrix effect results were both lower than 15%, showing that this method provided very clean extracts [27,28].

3. Liquid Chromatography Methods

HPLC is a conventional analytical method because of its low demand for instruments, and has been widely used in the analysis of kanamycin in different samples [36]. Depending on the retention mechanisms, the chromatographic columns used in this review were mainly divided into three types: Reversed-phase (RP) column, mixed-mode column, and hydrophilic interaction chromatography (HILIC) column. Each column type is marked in Tables 1–3. The different detectors coupled with HPLC mainly include UV/Fluorescence, Evaporative Light Scattering Detector (ELSD)/Pulsed Electrochemical Detection (PED), and Mass Spectrometry. The following content will be unfolded mainly on the basis of the classification above.

3.1. *UV and Fluorescence-Reserved Phase Liquid Chromatography after Derivatization*

Kanamycin is very polar and lacks chromophore or fluorophore, which makes it difficult to separate using traditionally reverse phase liquid chromatography (RPLC) recruiting UV or fluorescence monitoring. To overcome this problem, researchers have employed many different pre-column or post-column derivatization agents [47].

Derivatization of kanamycin is mainly focused on modifying its primary amine functions. The commonly used pre-column derivatization reagents include Phenylisocyanate (PIC) [48], 4-chloro-3,5-dinitrobenzotrifluoride(CNBF) [31], 1-naphthyl isothiocyanate (NITC) [13] and 9-fluorenylmethyl chloroformate (FMOC-Cl) [23]. Another reagent *o*-phthaldialdehyde (OPA) [35] can also be employed both in pre-column and post-column derivatization. Table 1 shows HPLC applications in the analysis of kanamycin with UV and fluorescence detection.

3.1.1. Pre-Column Derivatization

Pre-column derivatization of kanamycin changes its polarity, which optimizes its applicability for being analyzed through conventional RPLC. For example, CNBF was used as a pre-column derivatization reagent in kanamycin analysis in different kinds of soil samples with a UV detector at 245 nm with the reaction scheme as presented in Table 4 [31]. CNBF was able to react with primary and secondary amines in alkali condition, producing stable N-substituted-2, 6-dinitro-4-(trifluoromethyl)-benzamine derivative [49]. Unlike FOMC-Cl, CNBF does not need to be removed after derivatization. The analytical column was a kromasil C18 ODS column (250 × 4.6 mm, 5 μm). The SPE column was an AccuBOND ODS-C18 (3 mL/200 mg). Linearity range was 0.01–10.0 mg/kg, and LOD was 0.006 mg/kg. The HPLC-UV Chromatogram of CNBF-kanamycin A derivative is shown in Figure 3 [31].

Table 1. HPLC applications in the analysis of kanamycin A with UV and fluorescence detection.

Detection	Matrix	Compound of Interest	Derivatization Agent and Condition	Extraction and Cleanup Methods	HPLC Type	Column Type and Temperature	Mobile Phases	Detector Wave Length	Recovery (%)	LOD	LOQ	Refs
UV	solvent	KANA A	disodium tetraborate, added in mobile phase	-	IPLC	Reversed-phase column, XBridge C18 (250 × 4.6 mm, 5 µm), 50 °C	0.1 M disodium tetraborate (pH 9.0)-water (20:80, *v/v*) with 0.5 g/L sodium octanesulphonate, isocratic.	205 nm	-	38 mg/L	128 mg/L	[9]
UV	soil	KANA A	CNBF, PH 9.0, 10 min, 70 °C	**SPE**, AccuBOND ODS-C18	RPLC	Reversed-phase column, kromasil C18 (250 × 4.6 mm, 5 µm).	methanol-0.1% TFA in water, gradient	245 nm	72.3–92.5	0.006 mg/kg	0.01 mg/kg	[31]
UV	solvent	KANA A	PIC, 10 min, 70 °C	-	RPLC	Reversed-phase column, Phenomenex C18 (250 × 4.6 mm, 5 µm)	ACN-1% tris buffer (40:60, *v/v*) pH adjusted to 6.5 with 1 N sulfuric acid, isocratic	242 nm	92–98	0.597 µg/mL	1.021 µg/mL	[48]
UV	human plasma	KANA A	NITC in pyridine, 70 °C	**Protein Precipitation** with can	RPLC	Reversed-phase column, LichrocartPurospher STAR C18 (55 × 4 mm, 3 µm)	water-methanol (33:67, *v/v*), isocratic	230 nm	95.9–100.8	0.3 µg/mL	1.2 µg/mL	[13]
FL	animal feeds	KANA A	OPA-ME	**SPE**, Oasis MCX cartridge (3 cc, 60 mg)	RPLC	Reversed-phase column, XTerra™C18 (250 × 4.6 mm, 5 µm)	ammonium acetate solution-ACN (50:50, *v/v*), isocratic	Ex:230 nm; Em: 389 nm	98.4–106	5 g/ton	10 g/ton	[35]
FL	swine tissue	KANA A	FMOC-Cl, 15 min, RT	**SPE**, Oasis MCX cartridge (3 cc, 60 mg)	RPLC	Reversed-phase column, Waters symmetry C18 (150 × 3.9 mm, 5 µm)	ACN-water, gradient	Ex: 260 nm; Em: 315 nm	80.7–91.3	muscle: 0.03 mg/kg; liver: 0.06 mg/kg; kidney: 0.18 mg/kg	muscle: 0.1 mg/kg; liver: 0.2 mg/kg; kidney: 0.6 mg/kg.	[36]
FL	human plasma	KANA A	FMOC-Cl, 30 min, 25 °C	**Protein Precipitation** with ACN	RPLC	Reversed-phase column, Eclipse XDB C8 (150 × 4.6 mm, 5 µm), 25 °C	70% ACN, isocratic	Ex: 268 nm; Em: 318 nm	92.3–100.8	0.01 µg/mL	0.05 µg/mL	[12]
FL	wastewater and soil	KANA A	FMOC-Cl, 15 min, RT	**DLLME-SFO**, extraction solvent: dodecanol. dispersive solvent: ethanol	RPLC	Reversed-phase column, Diamonsil C18 (250 × 4.6 mm, 5 µm), 40 °C	ACN-water (84:16 *v/v*), isocratic	Ex: 265 nm; Em: 315 nm	Wastewater: 91.3–102.7; soil: 90.3–107.7	0.012 ng/mL	0.05 ng/mL	[23]
FL	pig feeds	KANA A	OPA-ME	**Protein Precipitation** with 0.1 M HCl	RPLC	Reversed-phase column, Eluent B: TSK ODS 120T (150 × 4.6 mm, 5 µm). Eluent C: Hypersil ODS for (150 × 3.2 mm, 5 µm), RT	Eluent B: THF-15 mM sodium sulphate, 3:97(*v/v*). Eluent C: 10 mM acetic acid-10 mM pentane sulphonate, 1.5:98.5 (*v/v*)	Ex: 355 nm; Em: 415 nm	89.4–92.8	0.2 mg/L	0.4 mg/L	[19]

FL: Fluorescence, KANA: Kanamycin, ACN: Acetonitrile.

Table 2. HPLC applications in the analysis of kanamycin with ELSD and PED.

Detection	Matrix	Compound of Interest	Extraction and Cleanup Methods	HPLC Type	Column Type and Temperature	Mobile Phases & IPR	Recovery (%)	LOD	LOQ	Refs
ELSD	solution	Kanamycin A, B and Sulfates	-	reversed-phase IPLC	Reversed-phase column, Spherisorb ODS-2 C18 (250 × 4.6 mm, 5 μm), RT	water-ACN (60:40, *v*/*v*), 11.6 mM HFBA, isocratic	Kanamycin A: 95–103, Kanamycin B: 95–105	Kanamycin A: 0.2 μg/mL, Kanamycin B: 1.4 μg/mL.	Kanamycin A: 0.6 μg/mL, Kanamycin B: 4 μg/mL	[50]
ELSD	solution	Kanamycin sulfate	-	HILIC	Mixed-mode column, Click TE-Cys (150 × 4.6 mm, 5 μm), 30 °C	ammonium formate aqueous solution (A)-ACN (B)-water(C), gradient	-	-	-	[51]
ELSD	fermentation broth	Kanamycin B	CD180 resin column	reversed-phase IPLC	Reversed-phase column, Agilent SB-Aq C18, RT	water-ACN (65:35, *v*/*v*), 11.6 mM HFBA, isocratic	93–96	-	0.05 mg/mL	[52]
PED	solution	Kanamycin	-	IPLC	Reversed-phase column, Platinum EPS (150 × 4.6 mm, 3 μm), 45 °C	**MPA**: sodium sulphate (5.0 g/L), sodium octanesulphonate (0.5 g/L) and 0.2M phosphate buffer pH 3.0 (50.0 mL/L) **MPB**: sodium sulphate (15 g/L), sodium octanesulphonate (0.5 g/L) and 0.2M phosphate buffer pH 3.0 (50.0 mL/L) Octanesulfonate as IPR, gradient	-	1.7 ng	5 ng	[10]
PED	solution	Kanamycin B,D	-	IPLC	Reversed-phase column, PLRP-S column packed with polystyrene-divinylbenzene, (250 × 4.6 mm, 8 μm), 45 °C	**MPA**: sodium sulphate (20 g/L), sodium octanesulphonate (1.3 g/L) and 0.2M phosphate buffer pH 3.0 (50.0 mL/L) **MPB**: sodium sulphate (60 g/L), sodium octanesulphonate (1.3 g/L) and 0.2M phosphate buffer pH 3.0 (50.0 mL/L). Octanesulfonate as IPR, gradient	-	Kanamycin D: 3 ng, Kanamycin B: 5 ng	Kanamycin D: 10 ng, Kanamycin B: 15 ng	[53]

ACN: Acetonitrile.

Table 3. HPLC applications in the analysis of kanamycin with MS detection.

Detection	Matrix	Compound of Interest	Extraction and Clean-up Methods	HPLC Type	Column Type and Temperature	Mobile Phases	Recovery (%)	LOD	LOQ	Refs
LC-MS/MS	rat plasma	kanamycin	**Protein Precipitation** with TCA added in sample	IPLC	Reversed-phase column, PhenomenexSynergi C12 Max-RP (50 × 2.0 mm, 4 μm)	0.1% FA in water-0.1% FA in ACN, gradient	-	-	20 ng/mL	[15]
UPLC-MS/MS TQD	bovine kidney, liver, muscle	kanamycin sulfate	**SPE**, Disposable pipet extraction (DPX), 5 mL tips containing 50 mg WCX sorbent	IPLC	Reversed-phase column, Waters BEH C18 (50 × 2.1 mm, 1.7 μm)	20 mM HFBA in 5% ACN in H_2O-20 mM HFBA in ACN, gradient, 3min. Tobramycin as IS	84–92	-	0.005 μg/g	[42]

Table 3. *Cont.*

Detection	Matrix	Compound of Interest	Extraction and Clean-up Methods	HPLC Type	Column Type and Temperature	Mobile Phases	Recovery (%)	LOD	LOQ	Refs
LC-MS/MS	muscle, kidney, liver, honey and milk	kanamycin	**SPE**, two-coupled Oasis HLB columns (3 mL/60 mg)	IPLC	Reversed-phase column, Capcell Pak C18 UG120 (150 × 2.0 mm, 5 μm), 30 °C	20 mM HFBA in 5% ACN-20 mM HFBA in 50% ACN, gradient,10 min. 500 ng/mL Tobramycin as IS	71–104	CCα (μg/kg) 49.5 for muscle, 48.9 for liver, 49.1 for kidney, 9.8 for honey, 121.5 for milk	CCβ (μg/kg) 60.2 for muscle, 58.4 for liver, 59.4 for kidney, 12.4 for honey, 146.4 for milk.	[33]
LC-MS/MS TQD	chicken meat	kanamycin	**Protein Precipitation** with ACN: 2% TCA (45:55, *v*/*v*). On-line clean-up using column: Thermo Cyclone P (50 × 0.5 mm, 60 μm)	IPLC	Mixed-mode column, Thermo Betasil phenyl hexyl (50 × 2.1 mm,3 μm), RT	1 mM HFBA and 0.5% FA in water-0.5% FA in ACN/methanol (1:1, *v*/*v*), gradient, 19 min	109–120	LOD: 10.0 μg/kg CCα: 121.3 μg/kg	LOQ: 25.0 μg/kg CCβ: 142.5 μg/kg	[21]
LC–MS/MS TQD	human serum	kanamycin	**Protein Precipitation** with TCA added in sample	IPLC	Reversed-phase column, Thermo Scientific™ HyPURITY™ C18 (5.0 × 2.1 mm, 3 μm)	water/methanol, containing 0.05% HFBA, gradient. Apramycin as IS	93.9–98.4	-	100 ng/mL	[17]
LC-MS-TOF	bovine milk & swine, poultry muscle	kanamycin	**Protein Precipitation** with TCA added in sample, then supernatant go through a tube containing bulk C18 resin	IPLC	Reversed-phase column, Waters X-Terras C18 (100 × 2.1 mm, 3.5 μm)	10 mM NFPA in H_2O-10 mM NFPA in ACN, gradient	milk: 92, muscle: 36.8–67	15 ng/g for milk and muscle	37.5 ng/g for milk, 25 ng/g for muscle	[20]
LC-MS/MS	muscle, liver, kidney, milk, egg	kanamycin	**SPE**, CBA cartridge (10 CC, 500 mg)	HILIC	HILIC column, CAPCELL PAK ST (150 × 2.0 mm, 4 μm), 30 °C	ACN-0.1%TFA, gradient	75–98	CCα (μg/kg) 8.1 for muscle, 8.5 for egg, 10.0 for liver, 11.2 for kidney, 11.5 for milk	CCβ (μg/kg) 17.6 for muscle, 21.9 for egg, 19.1 for liver, 17.4 for kidney, 18.5 for milk	[43]
LC-MS/MS	milk	kanamycin	Consecutive **SPE** of Sep-pak tC18 (6 mL/500 mg) and Oasis WCX (6 mL/500 mg), extraction with 3% TCA	HILIC	HILIC column, Click TE-Cys HILIC (150 × 3 mm, 3 μm)	1% FA in H_2O-1% FA in 80% ACN, both containing 30 mM ammonium formate, gradient, 10 min	69.9–77.9	6.1μg/kg	19.4μg/kg	[30]
UPLC-MS/MS	milk sample	Kanamycin acid salt	**SPE**, Supel, MISPE-Aminoglycoside cartridge (3 mL/50 mg)	HILIC	HILIC column, PhenomenexKinetex HILIC (100 × 2.1 mm, 1.7 μm), 35 °C	150 mM ammonium acetate in 1% FA(A)-ACN(B), gradient, 12 min	70–106	13.6 μg/kg	45.5 μg/kg	[28]
LC-MS/MS	human plasma	kanamycin	**Protein Precipitation** with acidified methanol using HCl	HILIC	HILIC column, Atlantis HILIC (150 × 2.1 mm, 3 μm), 35 °C	0.1% FA in water-0.1% FA in ACN, gradient, 9.0 min	91.2–93.4	-	1 μg/mL	[14]

Table 3. *Cont.*

Detection	Matrix	Compound of Interest	Extraction and Clean-up Methods	HPLC Type	Column Type and Temperature	Mobile Phases	Recovery (%)	LOD	LOQ	Refs
UPLC-MS/MS	human serum	kanamycin	**Protein Precipitation** with acidified methanol using HCl	HILIC	Reversed-phase column, Waters HSS T3 (50 × 2.1 mm, 1.8 μm), RT	10 mM ammonium formate in 0.1% FA-ACN in 0.1% FA, gradient, 3 min	-	0.5 μg/mL	2.5 μg/mL	[16]
UPLC-MS/MS	dried blood spots	kanamycin	**Protein Precipitation** with acidified methanol using HCl	HILIC	Reversed-phase column, Waters Acquity HSS T3 (50 × 2.1 mm, 1.8 μm)	10 mM ammonium formate in 0.1% FA-ACN in 0.1% FA, gradient, 4 min	-	0.3 μg/mL	5.0 μg/mL	[18]
Quattro Ultima LC-MS/MS	muscle and kidney	kanamycin	**SPE**, CBX cartridge (500 mg)	ZIC-HILIC	HILIC column, SeQuant ZIC-HILIC (100 × 2.1 mm, 5 μm), 32 °C	1% FA in 150 mM ammonium acetate-ACN, gradient, 19 min	81.1–104	18 ng/g	58 ng/g	[44]
LC–MS/MS	muscle, kidney and milk	kanamycin	**SPE**, Chromabond HR-X cartridge (6 mL/500 mg)	ZIC-HILIC	HILIC column, SeQuant ZIC-HILIC (100 × 2.1 mm, 5 μm), 30 °C	1% FA with 200 mM ammonium acetate in 5% ACN-ACN, gradient, 16 min	95–102	CCα (μg/kg) 118 for muscle, 2829 for kidney, 172 for milk	CCβ (μg/kg) 153 for muscle, 3401 for kidney, 215 for milk	[29]
Quattro Ultima UPLC-MS/MS	honey samples	Kanamycin A disulphatedihydrate	**SPE**, WCX cartridge, Accell Plus CM (6 mL/500 mg)	ZIC-HILIC	HILIC column, SeQuant ZIC-HILIC (150 × 2.1 mm, 3.5 μm), 40 °C	pH 4.5 with 125 mM ammonium formate-0.2% FA in ACN, gradient, 6 min. Amikacin as IS	68–112	8 μg/L	27 μg/L	[39]
Quattro Premier	kidney and honey	Kanamycin A disulphatedihydrate	**SPE**, WCX cartridge, Accell Plus CM (6 mL/500 mg)	ZIC-HILIC	HILIC column, SeQuant ZIC-HILIC (150 × 2.1 mm,3.5 μm), 40 °C	PH 4.5 with 175 mM ammonium formate-0.2% FA in ACN, gradient, 6 min. Amikacin as IS	82–105	CCα (μg/kg): 50 for honey, 2733 for kidney	CCβ (μg/kg) 67 for honey, 2965 for kidney. LOQ (μg/kg): 41 for honey, 85 for kidney.	[40]
UPLC-MS/MS	honey, milk and liver	kanamycin	**SPE**, Taurus WCX cartridge	ZIC-HILIC	Mixed-mode column, Obelisc R (100 × 2.1 mm, 5 μm), 40 °C	0.1% FA in water-0.1% FA in ACN, gradient, 8.0 min	58–96	LOD (μg/kg): 1 for honey, 1 for milk. CCα: 3 for honey, 172 for milk, 793 for liver	LOQ (μg/kg): 3 for honey, 5 for milk. CCβ: 5 for honey, 175 for milk 881 for liver	[41]
HILIC-MS/MS	honey, milk and pork	kanamycin disulfate salt	**SPE**, Supel MISPE-Aminoglycoside cartridge (3 mL/50 mg)	ZIC-HILIC	HILIC column, Zwitterionic HILIC (50 × 2.1 mm, 3.5 μm), 40 °C	175 mmol/L ammonium formate and 0.3% FA-methanol and 0.3% FA, gradient, 8 min	72.8–97	10 μg/kg for honey, 11 μg/kg for milk and pork	34 μg/kg for honey, 36 μg/kg for milk and pork	[27]
HILIC-MS/MS	human serum	kanamycin	**SPE**, Oasis MCX cartridge (30 mg)	ZIC-HILIC	HILIC column, SeQuant ZIC-HILIC (100 × 2.1 mm)	A (5/95/0.2, *v/v/v*) and B (95/5/0.2, *v/v/v*) each being a mixture of ACN: 2 mM ammonium acetate: FA, gradient	-	-	100 ng/mL	[37]
LC-MS/MS	veal muscle	kanamycin disulfate salt	**LLE**, defatting using hexane	ZIC-HILIC	HILIC column, ZIC-HILIC (50 × 2.1 mm, 5 μm)	0.4% formic acid inwater/ACN, gradient, 15 min	-	6 ng/g	-	[22]

Table 4. Reaction schemes of kanamycin with different derivatization regents.

Reaction Scheme	Refs
CNBF + kanamycin $\xrightarrow{OH^-}$ CNBF-kanamycin derivative (**CNBF**)	[31]
PIC: $-N{=}C{=}O + RNH_2 \longrightarrow -NH-C(=O)-NH-R$ (**PIC**; **N-aryl-N'-phenyl urea**)	[48]
NITC: $N{=}C{=}S + RNH_2 \longrightarrow NH-C(=S)-NHR$ (**NITC**)	[13]
FMOC-Cl + $HN(R_1)(R_2) \longrightarrow$ FMOC-$N(R_1)(R_2)$ + HCl (**FMOC-Cl**)	[12]

Table 4. *Cont.*

Reaction Scheme	Refs
OPA + RNH_2 + $HOCH_2CH_2SH$ ⟶ (isoindole with SCH_2CH_2OH, N–R) + H_2O	[36]
$B_4O_7^{2-}$ + 9 H_2O ⇌ $4B(OH)_4^-$ + $2H^+$ $B(OH)_4^-$ + (aminosugar with HO, HO, OH, NH_2) ⇌ (borate ester with HO, HO, OH, NH_2)	[9]

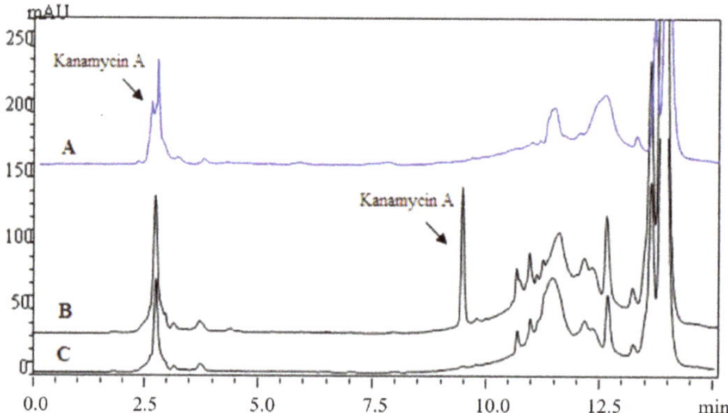

Figure 3. The HPLC-UV Chromatogram of CNBF-kanamycin A derivative. (A) The retention time of CNBF-kanamycin A derivative was 2.71 min without TFA in the mobile phase. The derivative could not be separated completely with interference. (B) The 0.1% TFA could improve separation efficiency. A perfect separation of CNBF-kanamycin A derivative was obtained with retention time of 9.58 min. (C) Blank soil sample.

PIC could react easily with primary or secondary amines, forming the stable *N*-aryl-*N*′-phenyl urea derivative, which was detected by UV at 242 nm. In Patel's study, a corresponding derivative through reaction of KANA with PIC (5 mg/mL in ACN) was formed in the presence of TEA for 10 min, followed by the RPLC method. The derivatives were separated on a Phenomenex C18 column (250 × 4.6 mm, 5 μm). Linearity range was 5–15 μg/mL. LOD was 0.597 μg/mL. The reaction scheme of PIC with kanamycin is presented in Table 4. The HPLC-UV Chromatogram of the kanamycin-PIC derivative is shown in Figure 4 [48].

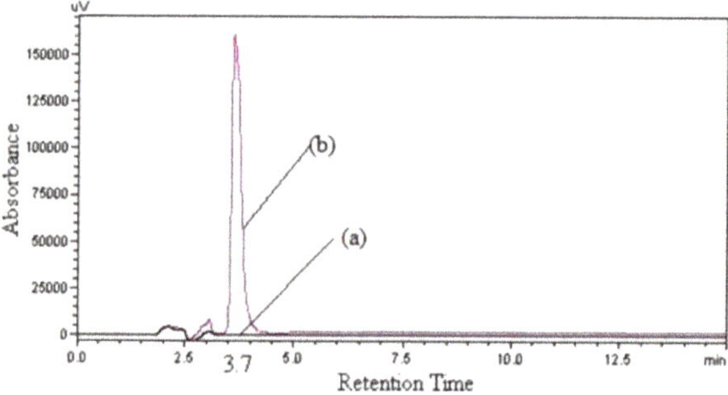

Figure 4. The HPLC-UV Chromatogram of the kanamycin-PIC derivative. (a) Blank; (b) Kanamycin-PIC derivative, 10 mg/mL showing retention time at 8.5 min.

NITC was used as a pre-column derivatization reagent to detect kanamycin A in human plasma by UV at 230 nm. The mixture containing kanamycin A was reacted in pyridine for 1 h. Methylamine was added to eliminate the remnant NITC after derivatization. The stationary phase was a Purospher STAR RP-18 column (55 × 4 mm, 3 μm). Linearity range was 1.2–40 μg/mL, and LOD was 0.3 μg/mL. The reaction scheme of NITC with kanamycin is presented in Table 4. The HPLC-UV Chromatogram of the kanamycin-NITC derivative is shown in Figure 5 [13].

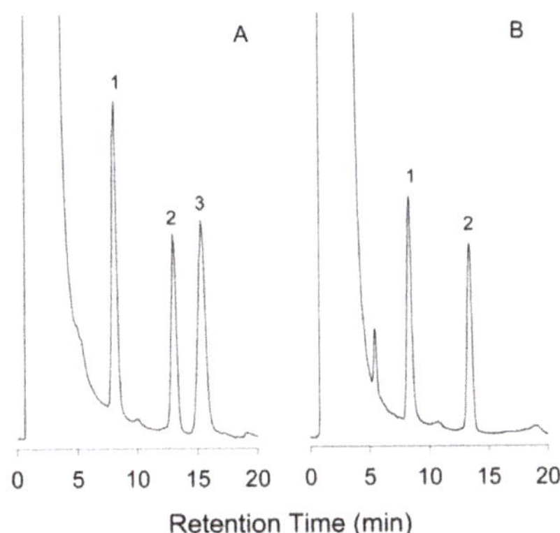

Figure 5. HPLC-UV chromatogram of the kanamycin-NITC derivative. (A) Separation of kanamycin A from kanamycin B, each at 40 µg/mL; (B) Determination of kanamycin A in commercial capsule sample. Peaks: 1, kanamycin A-NITC derivative; 2, acenaphthene (IS), 3, kanamycin B-NITC derivative.

FMOC-Cl was commonly used as a pre-column derivatization reagent of kanamycin, and the following detection was conducted by fluorescence. Kanamycin in human plasma reacted with FMOC-Cl in borate buffer solution (pH 8.5) for 30 min at room temperature, then separated by an Eclipse XDB C8 column (150 × 4.6 mm, 5 µm). LOD was 0.01 µg/mL, fluorescence wavelength was set at excitation of 268 nm and emission 318 nm. The reaction mechanism is shown in Table 4. The HPLC-FL Chromatogram of the kanamycin-FMOC derivative is shown in Figure 6 [12]. Similarly, pre-column FMOC-Cl derivatization of kanamycin was performed in swine tissue. The sample tissue was purified with the MCX SPE column. The derivatives were separated on a Waters symmetry C18 column (150 × 3.9 mm, 5 µm). LOD was 0.03 mg/kg for muscle, 0.06 mg/kg for liver and 0.18 mg/kg for kidney. The fluorescence measurements were set as excitation wavelength at 260 nm and emission wavelength at 315 nm. LOQ was 0.025 µg/mL, which was far lower than that reported by other researchers [36]. Another FMOC-Cl derivatization was prepared in wastewater and soil using a Diamonsil C18 column (250 × 4.6 mm, 5 µm). This is the first reported analysis that reduced the kanamycin derivative with the DLLME-SFO procedure. The fluorescence was measured at excitation wavelength 265 nm and emission wavelength 315 nm [23].

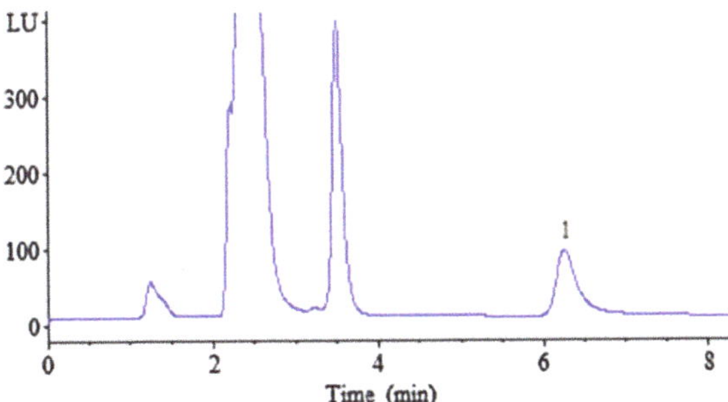

Figure 6. The HPLC-FL Chromatogram of the kanamycin-FMOC derivative. Kanamycin extracted from plasma from the same person 1.5 h after oral administration of 0.75 g of the drug. Peak 1, kanamycin-FMOC derivative.

OPA is a widely used derivatization reagent that introduces chromophores in HPLC methods using UV or fluorescence detection. A typical example is a pre-column derivatization of kanamycin with OPA in animal feeds; the reaction scheme is presented in Table 4 [35]. Oasis MCX SPE was used for cleanup. Chromatographic separation was implemented on a XTerra C18 column (250 × 4.6 mm, 5 μm). LOD was 5 g/ton in animal feeds with fluorescence measurement at excitation wavelength of 230 nm and emission wavelength of 389 nm. The HPLC-FL Chromatogram of the kanamycin-OPA pre-column derivative is shown in Figure 7 [35].

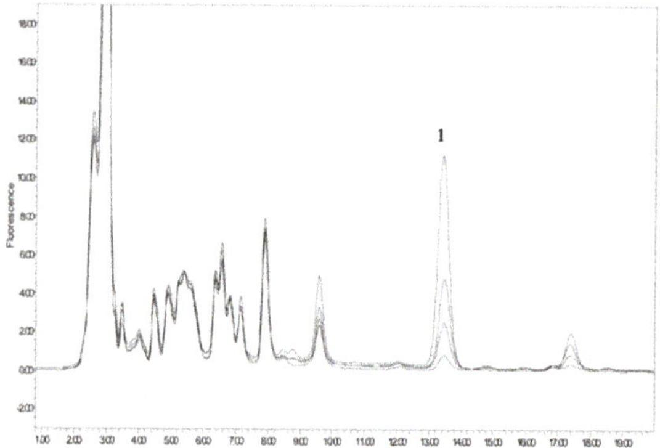

Figure 7. The HPLC-FL Chromatogram of kanamycin-OPA pre-column derivative. Peak 1: kanamycin-OPA derivative, with kanamycin in poultry feeds at levels of 10 mg/g, 40 mg/g, 80 mg/g, and 200 mg/g.

Although the pre-column derivatization methods can avoid using ion pair reagent (IPR), IPR is still needed under certain conditions. The derivatization of kanamycin using borate complexation is an example of this [9]; with reaction scheme is shown in Table 4. The HPLC-UV chromatogram of the kanamycin A-borate derivative is shown in Figure 8 [9]. Borate ion was obtained by dissolving borax in water. After borate complexation formation, the derivatives were analyzed with a XBridge C18 column (250 × 4.6 mm, 5 μm), using sodium octanesulphonate as IPR, and with UV detection at 205 nm. Baseline separation from kanamycins B, C, and D were achieved.

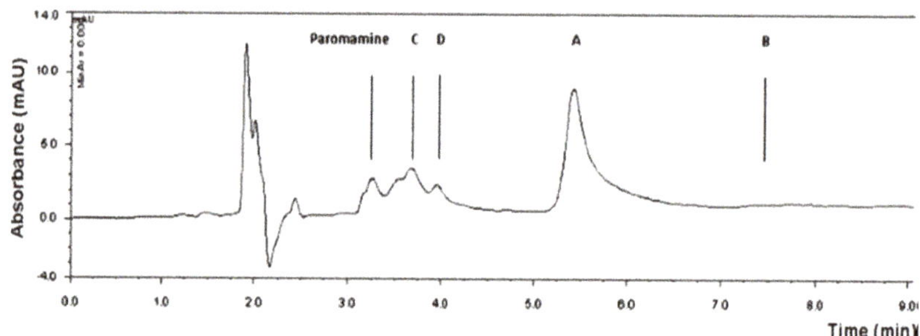

Figure 8. The HPLC-UV chromatogram of kanamycin A borate complexation. Chromatogram obtained after injection of kanamycin A solution (1 g/L) spiked with kanamycins B, C, and D, and paromamine (0.1 g/L each).

3.1.2. Post-Column Derivatization

Post-column derivatization requires more complicated instruments [47] and is confined by reaction time and the solvent system. However, the chemical reaction does not need to be complete since it is repeatable, and long-term stability of the derivative is not a concern [47].

OPA could be used as both pre-column [35] and post-column [19] derivatization agent. Post-column derivatization of kanamycin using OPA was achieved after RPLC with a C8 TSK ODS 120T (150 × 4.6 mm, 5 μm) or Hypersil ODS column (150 × 3.2 mm, 5 μm). Both columns led to good results. The HPLC-FL chromatogram of the kanamycin-OPA post-column derivative is shown in Figure 9 [19]. LOD was 0.2 mg/L in pig feeds, detected with fluorescence measurement at excitation wavelength of 355 nm and emission wavelength of 415 nm.

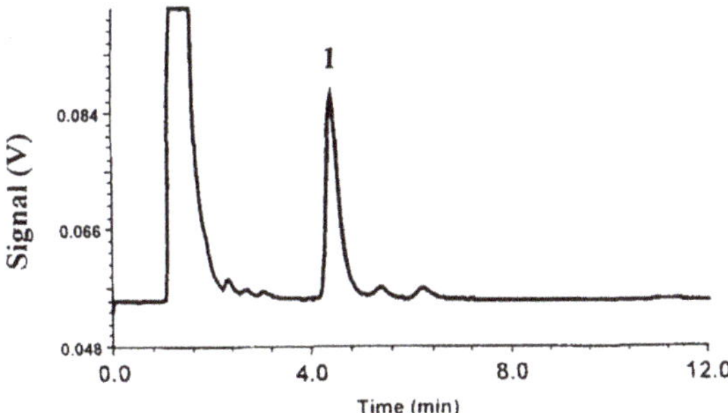

Figure 9. The HPLC-FL chromatogram of the kanamycin-OPA post-column derivative. Peak 1: kanamycin-OPA derivative, with kanamycin in swine feed at a level of 120 mg/kg.

3.2. ELSD and PED-Ion Pair Liquid Chromatography

In ion-pair liquid chromatography (IPLC) methods, the ion pairing reagent (IPR) is used as a mobile phase additive, which interacts with the RPLC stationary phase [47] and allows separating of the ionic and highly polar compounds on RP-HPLC columns. Alkyl sulfonates compounds like octanesulfonate could be used as IPR [10]. Meanwhile, volatile TFA and heptafluorobutyric acid (HFBA) [50,52] could also be used as IPR when coupled with MS detection. Since the high potency of IPR (>20 mM) is harmful to the column packing material, it is ideal to minimize the potency so as to achieve appropriate retention and peak shape [47].

In the IPLC method, an extra buffer system is required to maintain a stable pH of the mobile phase [47]. Ammonium acetate and phosphate are the most frequently used buffer solutions. Phosphate buffer is compatible with UV but not with an MS or ELSD detector. Meanwhile, ammonium acetate buffer is incompatible with UV but compatible with an MS detector [47].

3.2.1. Evaporative Light Scattering Detection (ELSD)

ELSD is increasingly being applied in IPLC for compounds without chromophores, because it eliminates the necessity of derivatization [50]. For HPLC applications in the analysis of kanamycin with ELSD detection, refer to Table 2. Some applications of the IPLC-ELSD methods are discussed hereinafter.

The separation of kanamycins A, B, and sulfate were validated through a novel IPLC-ELSD method without the derivatization step. Chromatographic separations were carried out with a Spherisorb ODS-2C18 column (250 × 4.6 mm, 5 μm) using 11.6 mM HFBA as IPR. The LODs were 0.20 μg/mL for kanamycin A, 1.4 μg/mL for kanamycin B and 2.3 μg/mL for kanamycin sulfates [50]. Another example of the IPLC-ELSD method was determination of kanamycin B and tobramycin impurities with HFBA as IPR. Kanamycin was separated on an Agilent SB-Aq C18 column (150 × 4.6 mm, 5 μm) after sample extraction on a weak acidic cation-exchange resin CD180 [52].

HILIC is a very important alternative approach for the separation of kanamycin. A new HILIC-coupled ELSD method was applied for kanamycin detection. In this research, a HILIC column Click TE-Cys (150 × 4.6 mm, 5 μm) was applied for selective separation of kanamycin. High buffer potency (≥50 mM) and low pH (2.7 or 3.0) are required for the mobile phase to improve peak shape and selectivity [51].

3.2.2. Pulsed Electrochemical Detection (PED)

HPLC together with pulsed electrochemical detector (PED) has been adopted in US Pharmacopoeia [50]. Analysis of kanamycin A and its related substances using IPLC coupled with PED has been reported [10,53]. For IPLC-ELSD applications in the analysis of kanamycin, refer to Table 2.

In Adams' work, octanesulfonate was selected as the IPR. To improve the sensitivity of PED detection, 0.5 M NaOH was added in the post-column effluent to adjust the pH to 13. The packing materials of column PLRP-S (250 × 4.6 mm, 8 μm) was poly (styrene-divinylbenzene). Eight components including kanamycin B and D were separated, and the method was applied to commercial samples [53].

Manyanga improved Adams' work [53] and applied the method to silica-based columns Platinum EPS (150 × 4.6 mm, 3 μm). The amount of salt in the mobile phase was reduced to improve stability, with the use of IPR of octanesulfonate remaining [10]. This method indicated better selectivity and sensitivity.

Nevertheless, the PED method has some disadvantages [54]. First, experience is important for repeatable quantitative results. Second, long equilibration time is required after washing of the electrodes of the electrochemical cell. Therefore, the PED method demands further improvement.

3.3. Liquid Chromatography-Mass Spectrometry

LC-MS/MS is a common analytical method in antibiotics residue analysis [33]. Applications of MS with RPLC, IPLC, HILIC or ZIC-HILIC in the analysis of kanamycin are discussed below; refer to Table 3. Mass spectral acquisition was performed in positive-ion mode by applying multiple reactions monitoring (MRM) using electrospray ionization (ESI) or atmospheric pressure chemical ionization (APCI) to detect kanamycin in this review. Kanamycin B produced $[M + H]^+$ ions at *m*/*z* 484, which is the precursor ion (Q1). The most abundant product ion (Q3) from the fragmentation was at *m*/*z* 324, and the relatively abundant product ions were *m*/*z* 205 and *m*/*z* 163. The three transition Q3 fragments of kanamycin were 163 for KANA1, and 324 or 205 for KANA2, respectively. The MS/MS spectra of kanamycin B was shown in Figure 10, and the fragmentation pathway of kanamycin B was shown in Figure 11 [55].

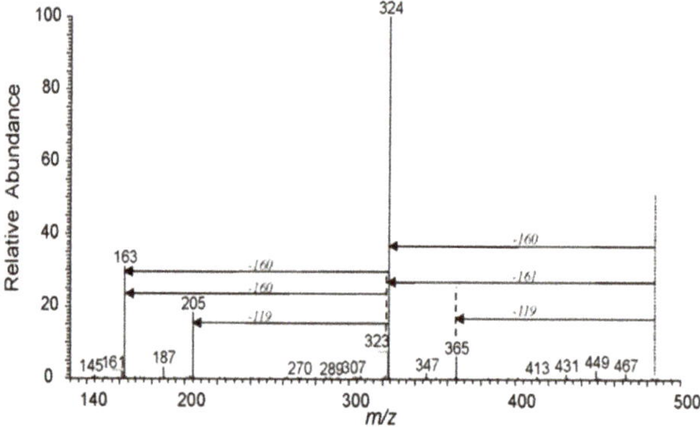

Figure 10. MS/MS spectra of $[M + H]^+$ ions of kanamycin B at *m*/*z* 484.

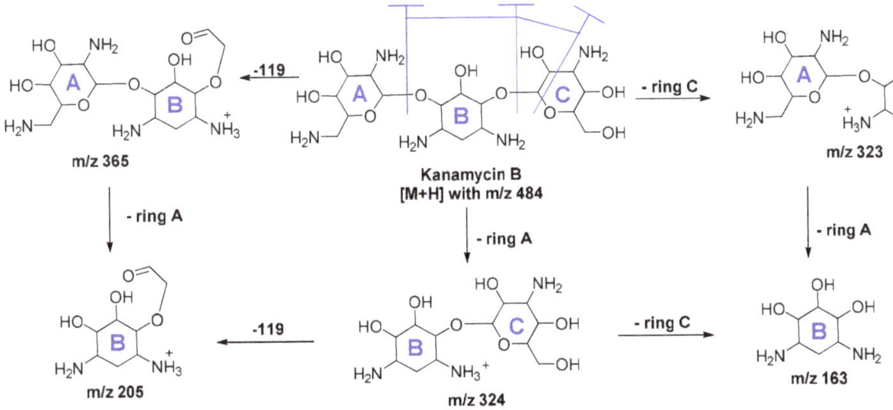

Figure 11. Summary of the fragmentation pathway of kanamycin B reference substances.

3.3.1. IPLC-MS/MS

The IPLC with MS/MS is a powerful tool commonly used in the separation of aminoglycosides [56,57]. The widely used IPRs in kanamycin IPLC-MS-MS analysis were HFBA [17,21,42], TCA [15] and Nonafluoropentanoic acid (NFPA) [20].

In a recent example, kanamycin along with other 12 aminoglycoside antibiotics (AGs) was determined in muscle, kidney, liver, honey, and milk [33]. Volatile HFBA was used as IPR, which was compatible with mass spectrometry and could cause strong retention on the reversed-phase column. Separation was performed using Capcell Pak C18 UG120 column (150 × 2.0 mm, 5 μm). Tobramycin was used as an internal standard (IS). Another rapid qualitative determination of 9 AGs including kanamycin in bovine matrix was realized by IPLC-MS/MS on a Waters BEH C18 (50 × 2.1 mm, 1.7 μm) column, using HFBA as IPR and tobramycin as IS. Since the column material particles were only 1.7 μm ID, the analysis time was shortened to 2.4 min [42]. In another multi-residue study, kanamycin together with 35 other antibiotics were detected in chicken meat on a Betasil phenyl hexyl column (50 × 2.1 mm, 3 μm) [21]. HFBA was chosen as an optimal IPR with minor or no ion suppression effect. For kanamycin detection, LOQ was 25 μg/kg, the decision limit CCα was 121 μg/kg, and detection capability CCβ was 143 μg/kg. Another example was the determination of kanamycin and amikacin in serum using IPLC-MS [17]. IPLC separation was achieved through a water-methanol gradient, containing 0.05% HFBA as IPR, on a Thermo ScientificTM HyPURITYTM C18 column (5.0 × 2.1 mm, 3 μm). Apramycin was used as IS solution [17].

Kanamycin, gentamicin and apramycin were quantified in rat plasma by Cheng et al. [15]. In this research, TCA acted as both a protein precipitator and an IPR, which only existed in the sample but not in the mobile phase; yet the system yielded better sensitivity. The absence of TCA in the mobile phase could reduce the contamination of ion source and result in good reproducibility [15]. The retention of AGs was improved on the Phenomenex Synergi C12 Max-RP column (50 × 2.0 mm, 4 μm), using tobramycin as the internal standard.

In a multi-residue analysis, kanamycin and nine other AGs were detected in bovine milk and bovine, swine and poultry muscle using a Waters X-Terra C18 column (100 × 2.1 mm, 3.5 μm) [20]. NFPA was used as IPR in the mobile phase, which improved kanamycin retention in the C18 column and improved its ionization, enhancing the MS/MS signal. Monitoring and screening was performed by LC-QTOF-MS and then confirmed by the LC-MS/MS method. LOQs for kanamycin were 37.5 ng/g for milk and 25 ng/g for muscle. The LODs for kanamycin was 15 ng/g in milk and muscle [20].

3.3.2. HILIC-MS/MS

HILIC shows a similar separation to normal phase liquid chromatography (NPLC), but it can also use water and volatile buffering solution as the mobile phases of RPLC, which are compatible with MS. Therefore, this technique can be applied to separate strong polar and hydrophilic chemical compounds [47].

Kanamycin is extremely hydrophilic because it has many amino and hydroxyl groups, so it has good solubility in the aqueous mobile phases of HILIC [58]. There is no need to use IPR in the mobile phase of HILIC, so it will cause less ion suppression and is fully compatible with MS systems. HILIC can provide higher sensitivity because the organic solvent-rich mobile phase is more volatile and can enhance desolvation and ionization efficiency of the ESI source [47].

3.3.3. ZIC-HILIC-MS/MS Method

In recent years, HILIC-coupled mass spectrometry has been successfully applied to the separation of AGs. The application of HILIC to quantify kanamycin and other 5 AGs in human serum was reported [37], with a zwitterionic ZIC-HILIC column (100 × 2.1 mm). LOQ of the method was 100 ng/mL for kanamycin [37].

Another application was reported in kidney and muscle tissues using a ZIC-HILIC column (100 × 2.1 mm, 5 μm) [44]. The LOQ of kanamycin was low—50 ng/g. It was observed that the high sorption affinity of kanamycin to polar surfaces required only polypropylene during sample preparation and storage, thus glass was avoided [44].

Kanamycin together with six other AGs was determined in veal muscle, and a ZIC-HILIC column (50 × 2.1 mm, 5 μm) was applied [22]. The ZIC-HILIC column (50 × 2.1 mm, 3.5 μm) was also used to determine kanamycin in honey, milk and pork samples [27].

Kumar et al. compared six kinds of HILIC stationary phases, including bare silica (anionic), amino phenol (cationic), amide (neutral), and zwitter ionic (ZIC) materials [39]. They concluded that the ZIC phase offered the best result, which might be attributed to the ZIC phase providing interaction with both the electropositive amino and the electronegative hydroxyl. The zwitterionic ZIC-HILIC column (150 × 2.1 mm, 3.5 μm) was used to determine Kanamycin A disulphate dihydrate in honey matrix. Amikacin was selected as the internal standard. The linearity range was 70–2000 μg/L. LOD and LOQ were 8 μg/L and 27 μg/L, respectively [39]. The year after that research, the above-mentioned method was improved and applied to the honey and kidney sample analysis for kanamycin, and validated according to Commission Decision 2002/657/EC. The CCα were 50 μg/kg for honey and 2733 μg/kg for kidney. LOQs were 41 μg/kg for honey and 85 μg/kg for kidney, respectively. The linearity was narrowed down to 70–495 μg/kg for honey and 200–4375 μg/kg for kidney [40].

In another similar study, kanamycin was detected in muscle, kidney (cattle and pig) and cow's milk using ZIC-HILIC column (100 × 2.1 mm, 5 μm) [29], and the internal standard tobramycin was used. The CCα ranges from 118 μg/kg to 2829 μg/kg, and the CCβ range from 153 μg/kg to 3401 μg/kg [29].

The usage of a new ZIC-HILIC column Obelisc R (100 × 2.1 mm, 5 μm) was also reported when detecting kanamycin in honey, milk and liver [41]. Obelisc R is a mixed-mode zwitterionic-type LiSC stationary phase, which has a similar structure to ZIC-HILIC column. However, Obelisc R is better than ZIC-HILIC because it has better sensitivity for AGs. The CCα ranges from 3 μg/kg to 793 μg/kg, and CCβ ranges from 5 μg/kg to 881 μg/kg [41].

3.3.4. Other HILIC-MS/MS Methods

The HILIC column CAPCELL PAK ST (150 × 2.0 mm, 4 μm) was applied in separation of 15 AGs residues including kanamycin in animal tissues, milk and eggs [43]. Measurement was carried out through a Thermo electron TSQ Quantum MS. The CCβ of kanamycin ranges from 17.4 μg/kg to 21.9 μg/kg, which was lower than the MRL defined by EU, USA and other countries [43].

In another analysis, kanamycin was separated through an Atlantis HILIC column (150 × 2.1 mm, 3 μm) [14], using apramycin as the internal standard. The calibration range was 100–2500 ng/mL for kanamycin in human plasma [14].

A new Click TE-Cys HILIC column (150 × 3 mm, 3 μm) was used to separate kanamycin in milk sample [30]. The LOD and LOQ were 6.1 μg/kg and 19.4 μg/kg, respectively, and the calibration range was 40 ng/mL to 4000 ng/mL [30].

The Phenomenex Kinetex HILIC column (100 × 2.1 mm, 1.7 μm) was applied to analyze kanamycin residues in different kinds of milk [28]. The LOD and LOQ were 13.6 μg/kg and 45.5 μg/kg, respectively, and the calibration range was 45.5 μg/k to 250 μg/kg kanamycin in milk [28].

Waters HSS T3 column (50 × 2.1 mm, 1.8 μm) was used to analyze kanamycin in serum, gentamicin as IS solution. The LOD and LOQ were 0.5 μg/mL and 2.5 μg/mL, respectively [16]. The LOD and LOQ were further expanded to 0.3 μg/mL and 5.0 μg/mL, respectively, and tested in dried blood spots (DBSs) samples in another study [18].

4. Conclusions

The extraction and clean-up methods play a very important role in the analysis of kanamycin. A series of information on methodologies for extraction and clean-up of kanamycin have been published. The extraction and clean-up methods for kanamycin have been applied to a variety of matrices, including animal feeds, liver and kidney tissues, and serum, among others. When the sample contains protein, as milk and serum, protein precipitation is an initial and key step. After protein precipitation, liquid-liquid extraction can be performed to remove fats by using hexane. SPE can be used to remove salts that might affect the ionization of the MS detector.

Much progress has been achieved in kanamycin detection. However, numerous problems still exist and need to be addressed. The UV and fluorescence derivatization methods are time consuming, and the reaction by-products often cause difficulties in quantitation. Therefore, simpler and direct detection methods are preferred, such as PED [10] and ELSD [50–52]. Nevertheless, ELSD is less sensitive than PED, needs to use volatile additives, and does not display a direct linear relation with the amount injected [10]. Some are semi-quantitative determination methods. LC-MS/MS methods can ensure good sensitivity and separation ability to detect kanamycin in animal-origin food [21,30]. However, the required instruments are not commonly available in many laboratories owing to their high cost. The IPLC is also suitable for MS-MS detector, while the IPR must be volatile and compatible with MS detector with low ionization suppression. Besides IPLC, HILIC is fully compatible with MS systems and free from IPR in the mobile phase. Meanwhile, the HILIC method can achieve lower detection limits [47]. Therefore, the HILIC-MS-MS offers further direction. Moreover, the MRLs of kanamycin residues defined by the EU Commission Decision is still not quite comprehensive, such as the absence of honey; thus, more sample materials needed to be included. We hope that this paper provides some help for kanamycin detection.

Author Contributions: X.Z. wrote the paper and organized data; J.W. revised the paper; Q.W. and L.L. collected relevant literature; Y.W. provided language modification; H.Y. provided the ideas about the paper and financial support.

Funding: The research was funded by the National Natural Science Foundation of China (31601536).

Acknowledgments: The research was supported by the Science and Technology Research Project from Hubei Provincial Education Department (Q20181321), the Yangtze Fund for Youth Teams of Science and Technology Innovation (2016cqt02).

Conflicts of Interest: The authors declare no conflict of interest.

Abbreviations

ACN	Acetonitrile
AGs	Aminoglycoside antibiotics
anti-TB drug	anti-Tuberculosis drug
APCI	Atmospheric Pressure Chemical Ionization
CBA	Carboxylic Acid
CBX	Carboxypropyl
CCα	Decision Limit
CCβ	Detection Capability
CNBF	4-chloro-3,5-dinitrobenzotrifluoride
DBSs	Dried Blood Spots
DLLME-SFO	Dispersive Liquid-Liquid Microextraction Based on Solidification of Floating Organic Droplet
DPX	Disposable (or Dispersive) Pipet Extraction
ELSD	Evaporative light scattering detection
ESI	Electrospray ionization
FA	Fomic acid
FL	Fluorescence
FOMC-Cl	9-Fluorenylmethyl chloroformate
HCl	Hydrochloricacid
HFBA	Heptafluorobutyric acid
HILIC	Hydrophilic interaction chromatography
HLB	Hydrophilic-lipophilic balanced
HPLC	High Performance Liquid Chromatography
IPLC	Ion-pair liquid Chromatography
IPR	Ion-pairing agent
IS	Internal standard
KANA	Kanamycin
LC-MS	Liquid chromatography-mass spectrometry methods
LLE	Liquid-liquid extraction
LOD	Limit of detection
LOQ	Limit of quantification
ME	Mercaptoethanol
MIPs	Molecularly imprinted polymers
MPA	Mobile phase A
MPB	Mobile phase B
MRLs	Maximum residue limits
MRM	Multiple reactions monitoring
MS	Mass Spectrometry
NITC	1-Naphthyl isothiocyanate
NFPA	Nonafluoropentanoic acid
NPLC	Normal Phase liquid Chromatography
ODS	Octadecyl silane
OPA	*O*-phthaladehyde
PED	Pulsed Electrochemical Detection
PIC	Phenylisocyanate
Q1	Precursor ion
Q3	Product ion
RPLC	Reverse phase liquid chromatography
SPE	Solid-phase extraction
TCA	Trichloroacetic acid
TFA	Trifluoroacetic acid
UV	Ultraviolet
ZIC-HILIC	Zwitter ionic-hydrophilic interaction chromatography

References

1. Durante-Mangoni, E.; Grammatikos, A.; Utili, R.; Falagas, M.E. Do we still need the aminoglycosides? *Int. J. Antimicrob. Ag.* **2009**, *33*, 201–205. [CrossRef]
2. Salimizand, H.; Zomorodi, A.R.; Mansury, D.; Khakshoor, M.; Azizi, O.; Khodaparast, S.; Baseri, Z.; Karami, P.; Zamanlou, S.; Farsiani, H.; et al. Diversity of aminoglycoside modifying enzymes and 16S rRNA methylases in Acinetobacter baumannii and Acinetobacter nosocomialis species in Iran; wide distribution of aadA1 and armA. *Infect. Genet. Evol.* **2018**, *66*, 195–199. [CrossRef]
3. Lee, J.H.; Kim, H.J.; Suh, M.W.; Ahn, S.C. Sustained Fos expression is observed in the developing brainstem auditory circuits of kanamycin-treated rats. *Neurosci. Lett.* **2011**, *505*, 98–103. [CrossRef]
4. Jiang, M.; Karasawa, T.; Steyger, P.S. Aminoglycoside-Induced Cochleotoxicity: A Review. *Front. Cell Neurosci.* **2017**, *11*, 308. [CrossRef] [PubMed]
5. Shavit, M.; Pokrovskaya, V.; Belakhov, V.; Baasov, T. Covalently linked kanamycin-Ciprofloxacin hybrid antibiotics as a tool to fight bacterial resistance. *Bioorgan. Med. Chem.* **2017**, *25*, 2917–2925. [CrossRef] [PubMed]
6. Jiang, Y.F.; Sun, D.W.; Pu, H.B.; Wei, Q.Y. Ultrasensitive analysis of kanamycin residue in milk by SERS-based aptasensor. *Talanta* **2019**, *197*, 151–158. [CrossRef] [PubMed]
7. European Commission. European Commission Decision (2002/657/EC) of 12 August 2002 implementing council directive 96/23/EC concerning the performance of analytical methods and interpretation of results. *Off. J. Eur. Communities* **2002**, *221*, 8–36.
8. Isoherranen, N.; Soback, S. Chromatographic methods for analysis of aminoglycoside antibiotics. *J. AOAC Int.* **1999**, *82*, 1017–1045.
9. Blanchaert, B.; Poderós Jorge, E.; Jankovics, P.; Adams, E.; Van Schepdael, A. Assay of Kanamycin A by HPLC with Direct UV Detection. *Chromatographia* **2013**, *76*, 1505–1512. [CrossRef]
10. Manyanga, V.; Dhulipalla, R.L.; Hoogmartens, J.; Adams, E. Improved liquid chromatographic method with pulsed electrochemical detection for the analysis of kanamycin. *J. Chromatogr. A* **2010**, *1217*, 3748–3753. [CrossRef]
11. Khan, S.M.; Miguel, E.M.; de Souza, C.F.; Silva, A.R.; Aucelioa, R.Q. Thioglycolic acid-CdTe quantum dots sensing and molecularly imprinted polymer based solid phase extraction for the determination of kanamycin in milk, vaccine and stream water samples. *Sensor. Actuat. B-Chem.* **2017**, *246*, 444–454. [CrossRef]
12. Wang, L.; Peng, J. LC Analysis of Kanamycin in Human Plasma, by Fluorescence Detection of the 9-Fluorenylmethyl Chloroformate Derivative. *Chromatographia* **2008**, *69*, 519–522. [CrossRef]
13. Chen, S.H.; Liang, Y.C.; Chou, Y.W. Analysis of kanamycin A in human plasma and in oral dosage form by derivatization with 1-naphthyl isothiocyanate and high-performance liquid chromatography. *J. Sep. Sci.* **2006**, *29*, 607–612. [CrossRef]
14. Kim, H.J.; Seo, K.A.; Kim, H.M.; Jeong, E.S.; Ghim, J.L.; Lee, S.H.; Lee, Y.M.; Kim, D.H.; Shin, J.G. Simple and accurate quantitative analysis of 20 anti-tuberculosis drugs in human plasma using liquid chromatography-electrospray ionization-tandem mass spectrometry. *J. Pharmaceut. Biomed. Anal.* **2015**, *102*, 9–16. [CrossRef] [PubMed]
15. Cheng, C.; Liu, S.; Xiao, D.; Hansel, S. The Application of Trichloroacetic Acid as an Ion Pairing Reagent in LC–MS–MS Method Development for Highly Polar Aminoglycoside Compounds. *Chromatographia* **2010**, *72*, 133–139. [CrossRef]
16. Han, M.; Jun, S.H.; Lee, J.H.; Park, K.U.; Song, J.; Song, S.H. Method for simultaneous analysis of nine second-line anti-tuberculosis drugs using UPLC-MS/MS. *J. Antimicrob. Chemoth.* **2013**, *68*, 2066–2073. [CrossRef] [PubMed]
17. Dijkstra, J.A.; Sturkenboom, M.G.; Kv, H.; Koster, R.A.; Greijdanus, B.; Alffenaar, J.W. Quantification of amikacin and kanamycin in serum using a simple and validated LC-MS/MS method. *Bioanalysis* **2014**, *6*, 2125–2133. [CrossRef]
18. Lee, K.; Jun, S.H.; Han, M.; Song, S.H.; Park, J.S.; Lee, J.H.; Park, K.U.; Song, J. Multiplex Assay of Second-Line Anti-Tuberculosis Drugs in Dried Blood Spots Using Ultra-Performance Liquid Chromatography-Tandem Mass Spectrometry. *Ann. Lab. Med.* **2016**, *36*, 489–493. [CrossRef] [PubMed]

19. Morovjan, G.C.; Csokan, P.P.; Nemeth-Konda, L. HPLC Determination of Colistin and Aminoglycoside Antibiotics in Feeds by Post-Column Derivatization and Fluorescence Detection. *Chromatographia* **1998**, *48*, 32–36. [CrossRef]
20. Arsand, J.B.; Jank, L.; Martins, M.T.; Hoff, R.B.; Barreto, F.; Pizzolato, T.M.; Sirtori, C. Determination of aminoglycoside residues in milk and muscle based on a simple and fast extraction procedure followed by liquid chromatography coupled to tandem mass spectrometry and time of flight mass spectrometry. *Talanta* **2016**, *154*, 38–45. [CrossRef]
21. Bousova, K.; Senyuva, H.; Mittendorf, K. Quantitative multi-residue method for determination antibiotics in chicken meat using turbulent flow chromatography coupled to liquid chromatography-tandem mass spectrometry. *J. Chromatogr. A* **2013**, *1274*, 19–27. [CrossRef] [PubMed]
22. Martos, P.A.; Jayasundara, F.; Dolbeer, J.; Jin, W.; Spilsbury, L.; Mitchell, M.; Varilla, C.; Shurmer, B. Multiclass, multiresidue drug analysis, including aminoglycosides, in animal tissue using liquid chromatography coupled to tandem mass spectrometry. *J. Agr. Food Chem.* **2010**, *58*, 5932–5944. [CrossRef]
23. Hu, S.; Song, Y.h.; Bai, X.H.; Jiang, X.; Yang, X.L. Derivatization and Solidification of Floating Dispersive Liquid-phase Microextraction for the Analysis of Kanamycin in Wastewater and Soil by HPLC with Fluorescence Detection. *CLEAN-Soil Air Water* **2014**, *42*, 364–370. [CrossRef]
24. Jia, X.Y.; Gong, D.R.; Xu, B.; Chi, Q.Q.; Zhang, X. Development of a novel, fast, sensitive method for chromium speciation in wastewater based on an organic polymer as solid phase extraction material combined with HPLC-ICP-MS. *Talanta* **2016**, *147*, 155–161. [CrossRef] [PubMed]
25. Zhou, Q.X.; Lei, M.; Li, J.; Liu, Y.L.; Zhao, K.F.; Zhao, D.C. Magnetic solid phase extraction of N- and S-containing polycyclic aromatic hydrocarbons at ppb levels by using a zerovalent iron nanoscale material modified with a metal organic framework of type Fe@MOF-5, and their determination by HPLC. *Microchim. Acta* **2017**, *184*, 1029–1036. [CrossRef]
26. Wang, T.; Zhu, Y.; Ma, J.; Xuan, R.; Gao, H.; Liang, Z.; Zhang, L.; Zhang, Y. Hydrophilic solid-phase extraction of melamine with ampholine-modified hybrid organic-inorganic silica material. *J. Sep. Sci.* **2015**, *38*, 87–92. [CrossRef] [PubMed]
27. Yang, B.; Wang, L.; Luo, C.; Wang, X.; Sun, C. Simultaneous Determination of 11 Aminoglycoside Residues in Honey, Milk, and Pork by Liquid Chromatography with Tandem Mass Spectrometry and Molecularly Imprinted Polymer Solid Phase Extraction. *J. AOAC Int.* **2017**, *100*, 1869–1878. [CrossRef] [PubMed]
28. Moreno-Gonzalez, D.; Hamed, A.M.; Garcia-Campana, A.M.; Gamiz-Gracia, L. Evaluation of hydrophilic interaction liquid chromatography-tandem mass spectrometry and extraction with molecularly imprinted polymers for determination of aminoglycosides in milk and milk-based functional foods. *Talanta* **2017**, *171*, 74–80. [CrossRef]
29. Bohm, D.A.; Stachel, C.S.; Gowik, P. Validation of a method for the determination of aminoglycosides in different matrices and species based on an in-house concept. *Food Addit. Contam. A* **2013**, *30*, 1037–1043. [CrossRef]
30. Wang, Y.; Li, S.; Zhang, F.; Lu, Y.; Yang, B.; Zhang, F.; Liang, X. Study of matrix effects for liquid chromatography-electrospray ionization tandem mass spectrometric analysis of 4 aminoglycosides residues in milk. *J. Chromatogr. A* **2016**, *1437*, 8–14. [CrossRef]
31. Sun, Y.; Li, D.; He, S.; Liu, P.; Hu, Q.; Cao, Y. Determination and dynamics of kanamycin A residue in soil by HPLC with SPE and precolumn derivatization. *Int. J. Environ. Anal. Chem.* **2013**, *93*, 472–481. [CrossRef]
32. Mahrouse, M.A. Simultaneous ultraperformance liquid chromatography/tandem mass spectrometry determination of four antihypertensive drugs in human plasma using hydrophilic-lipophilic balanced reversed-phase sorbents sample preparation protocol. *Biomed. Chromatogr.* **2018**, *32*, e4362. [CrossRef]
33. Zhu, W.X.; Yang, J.Z.; Wei, W.; Liu, Y.F.; Zhang, S.S. Simultaneous determination of 13 aminoglycoside residues in foods of animal origin by liquid chromatography-electrospray ionization tandem mass spectrometry with two consecutive solid-phase extraction steps. *J. Chromatogr. A* **2008**, *1207*, 29–37. [CrossRef] [PubMed]
34. Casado, N.; Damian, P.Q.; Morante-Zarcero, S.; Sierra, I. Bi-functionalized mesostructured silicas as reversed-phase/strong anion-exchange sorbents. Application to extraction of polyphenols prior to their quantitation by UHPLC with ion-trap mass spectrometry detection. *Microchim. Acta* **2019**, *186*, 1–13. [CrossRef] [PubMed]

35. Zhou, Y.X.; Yang, W.J.; Zhang, L.Y.; Wang, Z.Y. Determination of Kanamycin A in Animal Feeds by Solid Phase Extraction and High Performance Liquid Chromatography with Pre-Column Derivatization and Fluorescence Detection. *J. Liq. Chromatogr. R. T.* **2007**, *30*, 1603–1615. [CrossRef]
36. Chen, Y.; Chen, Q.; Tang, S.; Xiao, X. LC method for the analysis of kanamycin residue in swine tissues using derivatization with 9-fluorenylmethyl chloroformate. *J. Sep. Sci.* **2009**, *32*, 3620–3626. [CrossRef]
37. Oertel, R.; Neumeister, V.; Kirch, W. Hydrophilic interaction chromatography combined with tandem-mass spectrometry to determine six aminoglycosides in serum. *J. Chromatogr. A* **2004**, *1058*, 197–201. [CrossRef]
38. Tran, N.H.; Hu, J.Y.; Ong, S.L. Simultaneous determination of PPCPs, EDCs, and artificial sweeteners in environmental water samples using a single-step SPE coupled With HPLC-MS/MS and isotope dilution. *Talanta* **2013**, *113*, 82–92. [CrossRef]
39. Kumar, P.; Rubies, A.; Companyo, R.; Centrich, F. Hydrophilic interaction chromatography for the analysis of aminoglycosides. *J. Sep. Sci.* **2011**, *35*, 498–504. [CrossRef] [PubMed]
40. Kumar, P.; Rubies, A.; Companyo, R.; Centrich, F. Determination of aminoglycoside residues in kidney and honey samples by hydrophilic interaction chromatography-tandem mass spectrometry. *J. Sep. Sci.* **2012**, *35*, 2710–2717. [CrossRef]
41. Diez, C.; Guillarme, D.; Staub Sporri, A.; Cognard, E.; Ortelli, D.; Edder, P.; Rudaz, S. Aminoglycoside analysis in food of animal origin with a zwitterionic stationary phase and liquid chromatography-tandem mass spectrometry. *Anal. Chim. Acta* **2015**, *882*, 127–139. [CrossRef] [PubMed]
42. Lehotay, S.J.; Mastovska, K.; Lightfield, A.R.; Nunez, A.; Dutko, T.; Ng, C.; Bluhm, L. Rapid analysis of aminoglycoside antibiotics in bovine tissues using disposable pipette extraction and ultrahigh performance liquid chromatography-tandem mass spectrometry. *J. Chromatog. A* **2013**, *1313*, 103–112. [CrossRef]
43. Tao, Y.; Chen, D.; Yu, H.; Huang, L.; Liu, Z.; Cao, X.; Yan, C.; Pan, Y.; Liu, Z.; Yuan, Z. Simultaneous determination of 15 aminoglycoside(s) residues in animal derived foods by automated solid-phase extraction and liquid chromatography-tandem mass spectrometry. *Food Chem.* **2012**, *135*, 676–683. [CrossRef] [PubMed]
44. Ishii, R.; Horie, M.; Chan, W.; MacNeil, J. Multi-residue quantitation of aminoglycoside antibiotics in kidney and meat by liquid chromatography with tandem mass spectrometry. *Food Addit. Contam. A* **2008**, *25*, 1509–1519. [CrossRef] [PubMed]
45. Figueiredo, L.; Erny, G.L.; Santos, L.; Alves, A. Applications of molecularly imprinted polymers to the analysis and removal of personal care products: A review. *Talanta* **2016**, *146*, 754–765. [CrossRef]
46. Tan, F.; Sun, D.; Gao, J.; Zhao, Q.; Wang, X.; Teng, F.; Quan, X.; Chen, J. Preparation of molecularly imprinted polymer nanoparticles for selective removal of fluoroquinolone antibiotics in aqueous solution. *J. Hazard. Mater.* **2013**, *244*, 750–757. [CrossRef]
47. Farouk, F.; Azzazy, H.M.; Niessen, W.M. Challenges in the determination of aminoglycoside antibiotics, a review. *Anal. Chim. Acta* **2015**, *890*, 21–43. [CrossRef]
48. Patel, K.N.; Limgavkar, R.S.; Raval, H.G.; Patel, K.G.; Gandhi, T.R. High-Performance Liquid Chromatographic Determination of Cefalexin Monohydrate and Kanamycin Monosulfate with Precolumn Derivatization. *J. Liq. Chromatogr. R. T.* **2015**, *38*, 716–721. [CrossRef]
49. Qian, K.; Tao, T.; Shi, T.; Fang, W.; Li, J.; Cao, Y. Residue determination of glyphosate in environmental water samples with high-performance liquid chromatography and UV detection after derivatization with 4-chloro-3,5-dinitrobenzotrifluoride. *Anal. Chim. Acta* **2009**, *635*, 222–226. [CrossRef]
50. Megoulas, N.C.; Koupparis, M.A. Direct determination of kanamycin in raw materials, veterinary formulation and culture media using a novel liquid chromatography–evaporative light scattering method. *Anal. Chim. Acta* **2005**, *547*, 64–72. [CrossRef]
51. Wei, J.; Shen, A.; Wan, H.; Yan, J.; Yang, B.; Guo, Z.; Zhang, F.; Liang, X. Highly selective separation of aminoglycoside antibiotics on a zwitterionic Click TE-Cys column. *J. Sep. Sci.* **2014**, *37*, 1781–1787. [CrossRef]
52. Zhang, Y.; He, H.M.; Zhang, J.; Liu, F.J.; Li, C.; Wang, B.W.; Qiao, R.Z. HPLC-ELSD determination of kanamycin B in the presence of kanamycin A in fermentation broth. *Biomed. Chromatogr.* **2015**, *29*, 396–401. [CrossRef]
53. Adams, E.; Dalle, J.; De, B.E.; De, S.I.; Roets, E.; Hoogmartens, J. Analysis of kanamycin sulfate by liquid chromatography with pulsed electrochemical detection. *J. Chromatogr. A* **1997**, *766*, 133–139. [CrossRef]
54. Chopra, S.; Vanderheyden, G.; Hoogmartens, J.; Van Schepdael, A.; Adams, E. Comparative study on the analytical performance of different detectors for the liquid chromatographic analysis of tobramycin. *J. Pharmaceut. Biomed. Anal.* **2010**, *53*, 151–157. [CrossRef]

55. Li, B.; Van Schepdael, A.; Hoogmartens, J.; Adams, E. Characterization of impurities in tobramycin by liquid chromatography-mass spectrometry. *J. Chromatogr. A* **2009**, *1216*, 3941–3945. [CrossRef]
56. Gremilogianni, A.M.; Megoulas, N.C.; Koupparis, M.A. Hydrophilic interaction vs ion pair liquid chromatography for the determination of streptomycin and dihydrostreptomycin residues in milk based on mass spectrometric detection. *J. Chromatogr. A* **2010**, *1217*, 6646–6651. [CrossRef]
57. Babin, Y.; Fortier, S. A high-throughput analytical method for determination of aminoglycosides in veal tissues by liquid chromatography/tandem mass spectrometry with automated cleanup. *J. AOAC Int.* **2007**, *90*, 1418–1426.
58. Buszewski, B.; Noga, S. Hydrophilic interaction liquid chromatography (HILIC)—A powerful separation technique. *Anal. Bioanal. Chem.* **2012**, *402*, 231–247. [CrossRef]

Review

An Update on Isolation Methods for Proteomic Studies of Extracellular Vesicles in Biofluids

Jing Li [1], Xianqing He [1], Yuanyuan Deng [2] and Chenxi Yang [2,*]

1 School of Chemistry, Chemical Engineering and Life Science, Wuhan University of Technology, 122 luoshilu, Wuhan 430070, China; lij@whut.edu.cn (J.L.); 714769257@whut.edu.cn (X.H.)
2 School of Biological Science & Medical Engineering, Southeast University, No.2 Sipailou, Nanjing 210096, China; 213150742@seu.edu.cn
* Correspondence: yangchenxi@seu.edu.cn

Academic Editors: Marcello Locatelli, Angela Tartaglia, Dora Melucci, Abuzar Kabir, Halil Ibrahim Ulusoy and Victoria Samanidou
Received: 10 September 2019; Accepted: 26 September 2019; Published: 27 September 2019

Abstract: Extracellular vesicles (EVs) are lipid bilayer enclosed particles which present in almost all types of biofluids and contain specific proteins, lipids, and RNA. Increasing evidence has demonstrated the tremendous clinical potential of EVs as diagnostic and therapeutic tools, especially in biofluids, since they can be detected without invasive surgery. With the advanced mass spectrometry (MS), it is possible to decipher the protein content of EVs under different physiological and pathological conditions. Therefore, MS-based EV proteomic studies have grown rapidly in the past decade for biomarker discovery. This review focuses on the studies that isolate EVs from different biofluids and contain MS-based proteomic analysis. Literature published in the past decade (2009.1–2019.7) were selected and summarized with emphasis on isolation methods of EVs and MS analysis strategies, with the aim to give an overview of MS-based EV proteomic studies and provide a reference for future research.

Keywords: extracellular vesicles; isolation methods; biofluid; proteomics; mass spectrometry

---

1. Introduction

Although extracellular vesicles (EVs) were first described as 'platelet dust' in the late 1960s, it is now widely accepted that EVs are novel and important mediators for cellular communication by delivering bioactive molecules from donor to recipient cells [1,2]. Growing evidence has indicated that the cargo of EVs can reflect the content of their cells of origin and regulate physiological and pathological processes [3]. To date, EVs are considered as a novel source for biomarker discovery. With the benefits of liquid biopsy, analysis of EVs in biofluids has emerged as a promising diagnostic and monitoring tool for many diseases including cancer, neurodegenerative, kidney, and cardiovascular diseases [1,4,5].

EVs are membrane-enclosed particles that carry many bioactive molecules, including nucleic acids, proteins, and lipids, from their cells of origin. Based on their intracellular origin, EVs can be classified into three categories: exosomes, microvesicles (MVs), and apoptotic bodies. Exosomes are classically defined as the nanoparticles with sizes from 30–100 nm and formed by the fusion of multivesicular bodies with the plasma membranes; microvesicles, also called ectosomes, are usually described as the particles with sizes from 100–1000 nm and directly budded from the plasma membrane; apoptotic bodies (>1000 nm) are often considered as the particles that are released by apoptotic cells [6,7]. Despite apparent differences from their definition, it is difficult to differentiate the types of EVs after their release. It has been shown that the size of exosomes and microvesicles has a considerable overlap [7]. Currently, most of the isolation methods described in this review result in the mixed population of EVs. In addition to the physical heterogeneity, EVs are also highly heterogeneous in their cargo composition. Significant efforts have been made with the aim to comprehensively categorize EV subtypes, such as building an extensive and up–to–date database for EVs including ExoCarta, Vesiclepedia, and EVpedia [8–11]. However, consensus regarding the molecular markers to unambiguously distinguish the types of EVs remains to be a problem. Therefore, 'extracellular vesicle', which is suggested by the International Society of Extracellular Vesicles (ISEV), is used here for all the secreted vesicles [12].

Due to their tremendously diagnostic and therapeutic potential, EVs have gained increasing attention in the past decade, as shown by the number of publications (Figure 1). However, most of the studies focus on the nucleic acid content of EVs, such as microRNA or messenger RNA. With its improvements on sensitivity and high-throughput, mass spectrometry (MS) has become the fundamental technique of proteomics in recent years. Nowadays, MS has the capability to identify and characterize the protein content of EVs [6]. In the past decades, MS has been utilized to study EV proteome in various diseases, such as cancer and cardiovascular diseases [13,14]. This review will focus on publications within ten years that contain MS-based studies for EV proteins in human biofluids, such as urine, plasma, and saliva, rather than studies of EVs from laboratory animals or cell cultures and without any MS characterization. The references may be not comprehensive, but we try to highlight the recent improvements on isolation and MS strategies used in studies of EV proteome.

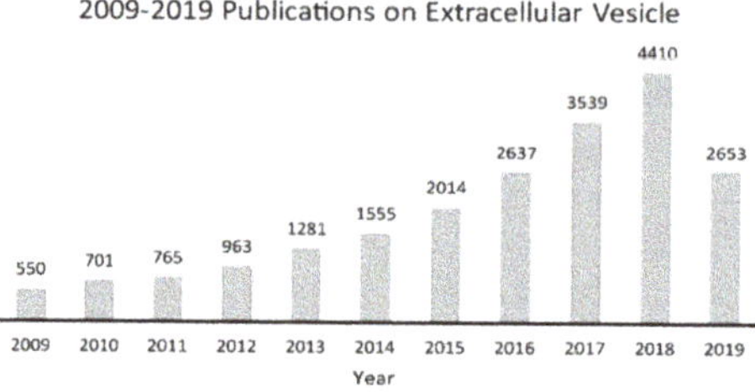

Figure 1. Publication trends on extracellular vesicle studies in the past decade (2009.1 to 2019.7). Publications were selected by searching the keyword "extracellular vesicle" in the Web of Science from the year of 2009.1 to 2019.7. x axis: year; y axis: number of publications.

2. Isolation Strategies for Extracellular Vesicles in MS-Based Proteomic Studies

EVs in biofluids are several orders of magnitude lower than other abundant components, such as lipoprotein particles, protein aggregates, and soluble proteins, including albumin in blood and Tamm-Horsfall protein (THP) in urine, which could interfere with the characterization of EVs [15,16]. Thus, the isolation step is required for all EV studies. In a typical MS-based bottom-up proteomic workflow, an additional isolation step for EVs is applied before the protein extraction and digestion (Figure 2). The commonly used isolation methods are either through the physical property of EVs, such as density and size, or based on the chemical property of EVs, such as through interacting with surface proteins of EVs, to achieve isolation [15]. Even though microfluidics-based devices hold promising potential for rapid and efficient isolation of EVs from biofluids, their low processing capacity greatly limits the downstream analysis due to the insufficient amounts of proteins [17]. Hence, this review will discuss the isolation methods, which could provide successful downstream MS-based proteomic EV studies and give an update for the ten-year improvements on isolation methods which are used in MS-based workflow studies.

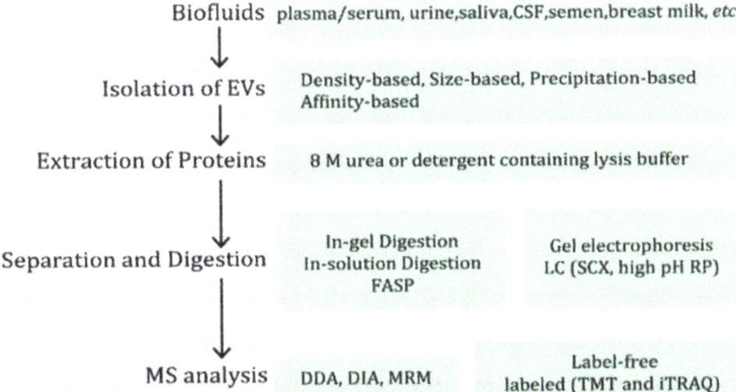

Figure 2. A general workflow of mass spectrometry (MS)-based proteomic extracellular vesicle (EV) study. EVs are firstly isolated from various biofluids, and EV proteins are extracted by adding detergent or non-detergent containing lysis buffer. The extracted EV proteins can be separated by gel electrophoresis and digested in-gel before MS analysis. Alternatively, digestion can be performed after protein extraction, and the generated peptides are either fractionated by liquid chromatography (LC) before MS analysis or directly subjected to MS analysis. The MS analysis can be conducted in data-dependent acquisition (DDA) or data-independent acquisition (DIA) for discovery EV studies or multiple reaction monitoring (MRM) for target EV studies. Differential expressed EV proteins also can be revealed by quantitative MS analysis via label-free or labeled quantitative proteomics. CSF: cerebrospinal fluid; FASP: filter aided sample preparation; SCX: strong cation exchange chromatography; RP: reverse phase chromatography; TMT: tandem mass tag; iTRAQ: isobaric tag for relative and absolute quantitation.

2.1. Sample Storage and Processing Conditions

Inappropriate storing and processing conditions can significantly affect the EV characteristics and recovery from biofluids, thus increasing pre-analytical variances or bringing artificial results. However, this aspect is not the focus of this review, and several comprehensive review or research papers have covered this topic [11,15,18–20]. Herein, some suggestions which are important and have been universally understood by the community are listed. In general, samples should be processed immediately after collection and in minimal waiting periods between each processing stages. Aliquots of samples are recommended in order to avoid multiple freezing–thawing cycles during whole processes. To obtain better EV recovery and preserve their characteristics in the biofluids, storing samples at −80 °C before EV isolation is important for long time storage [18,21–23]. However, one should be aware that there are no strict standards regarding sample storage and processing conditions for now. Most studies focus on the effects on concentration, size, RNA content, or some of the marker proteins of EVs under different conditions [18,21,24]. The comprehensive proteomic studies are still needed for evaluating the effects on protein content. In addition, each type of biofluid has special considerations which should be noticed before starting experiments.

2.2. Density-Based Isolation

Differential ultracentrifugation (dUC) as the current gold standard is the most commonly used isolation method of EVs. A recent worldwide survey of ISEV members has reported that 80% of EV isolation was conducted by dUC [25]. Biofluids typically contain a multicomponent mixture of particles that differ in sizes and densities, thus resulting in different sedimentation rates. During dUC, smaller particles can be isolated from larger ones according to their sedimentation rates by a successive increase of centrifugation forces and durations [26]. Although the details of protocols used by different groups are different to some extent, the general steps should be similar which usually include consecutively pelleting the apoptotic bodies and cell debris, the MVs, and the exosomes, as shown in Figure 3.

In most cases, samples are usually diluted by phosphate-buffered saline (PBS) before centrifugation to decrease their viscosity [27]. This dilution not only can increase the purity of EVs by decreasing the co-isolated contaminants, such as protein aggregates, but also can improve the efficiency of EV isolation since higher viscosity resulted in lower sedimentation efficiency [16,18,28]. After dilution, one or more centrifugation steps at 1000–3000× *g* are applied to remove dead cells and cell debris [15]. For example, a 30 min centrifugation at 2000× *g* can be used for viscous fluids according to one of the most cited protocols from Théry et al. [27].

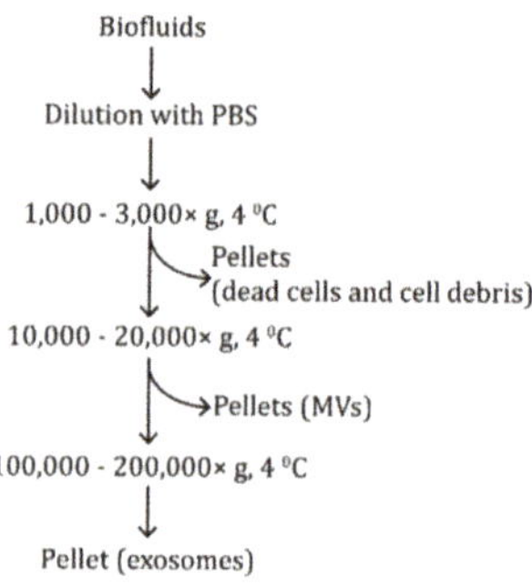

Figure 3. A basic differential ultracentrifugation (dUC) workflow for isolation of MVs and exosomes. Biofluids are diluted by phosphate-buffered saline (PBS) before centrifugation. Dead cells and cell debris are removed as pellets during the centrifugation at 1000–3000× *g*. Further centrifugation of supernatant at 10,000–20,000× *g* facilitates the isolation of MVs from exosomes. Finally, the recovery of exosomes is achieved by ultracentrifuging the 10,000–20,000× *g*-derived supernatant at 100,000–200,000× *g*.

Afterward, higher speed centrifugation, such as 10,000–20,000× *g*, typically follows to isolate MVs in the biofluids (Figure 3) [29,30]. The so-called ultracentrifugation at 100,000–200,000× *g* for hours is normally used to isolate exosomes from samples (Figure 3) [15,31]. Chutipongtanate et al. collected urinary MVs at a 20 min centrifugation of 10,000× *g* before proceeding to prepare urinary exosomes at 100,000× *g* for 1 h [32]. Sun et al. also isolated MVs and exosomes from saliva samples by sequentially centrifuging at 10,000× or 20,000× *g* for 1 h and 100,000× or 125,000× *g* for 2.5 h, with 785 proteins identified from MVs and 910 proteins from exosomes [33]. Table 1 lists the details of centrifugation force and time from the selected EV studies for future reference. Their corresponding MS strategies and results are also included in Table 1. Rather than using common gel-based bottom-up proteomics, different methodologies on MS-based workflow were also developed and applied to EV studies as summarized in Table 1, such as different liquid chromatography (LC) fractionation methods, digestion strategies, and MS acquisition approaches, which will be discussed in Section 4. Many exosomes studies discarded the pellets resulted from 10,000–20,000× *g* before ultracentrifugation at 100,000–200,000× *g* (Table 1). However, Whitham et al. recently isolated EVs at 20,000× *g* for 1 h to study the exercise-induced EV proteome and found that a host of small-vesicle and exosomal markers, such as SDCBP, TSG101, PDCD6IP (ALIX), CD63, and CD9, identified in 20,000× *g*-derived EV lysates. Further quantitative studies revealed that no significant differences were observed in any EV markers between samples subjected to 20,000× or 100,000× *g* centrifugation. They claimed that a quantitative proteomic analysis of small-vesicle and exosomal protein cargo was possible with the 20,000× *g* centrifugation for 1 h rather than prolonged centrifugation at 100,000× *g* [34]. Besides, Kim et al. claimed that centrifugation at 40,000× *g* could provide comparable or improved results relative to ultracentrifugation at 110,000× *g* [35]. Those studies may imply that the purity of exosome samples yielded by dUC are obtained with the cost of exosome loss during centrifugation at 10,000–20,000× *g*.

The pellets of interest are usually washed once at the final steps by resuspension and centrifugationagain. It has been demonstrated that less washing can result in a higher EV yield, but also have more contaminants [36]. Therefore, the balance between yield and purity should be judged when adopting protocols. It is also worth noting that the efficiency of isolation is not only dependent on the viscosity of the samples, centrifugation force, and time, but also on rotor type since sedimentation path lengths are dependent on the type of rotors used and different distances from the rotational axis could result in differences in the g-force. Cvjetkovic et al. applied a 70 min centrifugation at 100,000× *g* for exosome isolation on three different rotors and found that the yield and purity of exosomes obtained were significantly different [37]. To address this issue, a web-calculator was developed by Livshits et al. to adjust the common dUC protocol to the "individual" dUC protocol [26]. Therefore, one should be aware that proper modifications are necessary when adopting dUC for different types of biofluids and laboratory settings in order to achieve optimal isolation.

Table 1. Selected MS analysis for EVs obtained from centrifugation-based isolation.

Isolation	Proteomic Sample Preparation	Mass Spectrometry	Sample Origin	Number of Proteins	Year	Study
19,000× *g* for 120 min	2D-LC/MS: SCX as 1st dimensional	LTQ ion trap	plasma	1806 proteins	2017	[30]
Sucrose cushion at 100,000× *g* for 90 min	2D-LC/MS: C18-SCX stage-tip as 1st dimensional	Q-Exactive	serum	702 proteins	2017	[38]
100,000× *g* for 90 minincubation with DTT	iTRAQ 2D-LC/MS	LTQ-Orbitrap Velos Elite	urine	4710 proteins in total and 3528 proteins for quantification	2017	[39]
Sucrose cushion at 100,000× *g* for 90 min	iTRAQ 2D-LC/MS: high pH as 1st dimensional	Orbitrap Fusion Lumos	semen	3699 proteins in total	2018	[40]
110,000× *g* for 70 min	FASP	Q Exactive	serum	655 proteins	2018	[41]
10,000× *g*, 20 min for MVs and at 100,000× *g*, 1 h for exosomes	in-solution digestion	SWATH-MS TripleTof 5600+	urine	Targeted data analysis for 888 proteins	2018	[32]
Density ultracentrifugation at 270,000× *g*, 1 h and incubation with DTT	in-solution digestion	MS^E	urine	1877 proteins	2011	[42]
100,000× *g* for 180 min	in-solution digestion	L Q-Exactive Orbitrap	umbilical cord blood	211 proteins	2015	[43]
200,000× *g*, 1 h and incubation with DTT	in-gel digestion	LTQ Orbitrap XL and LTQ Orbitrap Velos	urine	1989 proteins in total	2012	[44]
100,000× *g* for 90 min	in-solution digestion	LTQ Orbitrap Velos	saliva	381 proteins	2015	[45]
200,000× *g* for 90 min and incubation with KBr	iTRAQ LC off-line separation	MALDI * tandem mass spectrometry	plasma	not report	2010	[46]
Sucrose cushion at 192,000× *g* for 15–18 h	in-gel digestion	Q-Exactive	breast milk	1963 proteins	2016	[47]
20,000× *g* for 1 h for MVs	in-solution digestion	Q-Exactive/Plus	plasma	3294 proteins in 4 h LC/MS	2015	[29]
10,000 or 20,000× *g*, 1 h for MVs; 100,000 or 125,000× *g*, 2.5 h for exosomes	SDS-PAGE FASP	Q-Exactive	saliva	785 proteins for MVs; 910 proteins for exosomes	2018	[33]
20,000× *g*, 1 h for MVs; 100,000× *g*, 1 h for exosomes	in-solution digestion	LTQ-Orbitrap Velos Pro	plasma	9225 phosphopeptides in MVs; 1014 phosphopeptides in exosomes	2017	[48]
100,000× *g* for 70 min	in-gel digestion	LTQ-XL	CSF	91 proteins identified from control466 proteins identified from disease	2018	[49]

* MALDI: Matrix-assisted laser desorption/ionization.

dUC has been utilized to isolate MVs and exosomes from different types of biofluids, such as plasma, urine, saliva, breast milk, and semen, as listed in Table 1. But the EV pellets obtained from dUC are usually contaminated with some co-sediment high abundant components in the biofluids including lipoprotein participles, protein aggregates, and high abundant soluble proteins, which significantly affect the downstream MS analysis. To improve the purity of isolated EVs, density gradient (DG) flotation, such as the sucrose gradient or OptiPrep velocity gradient (iodixanol gradient), is developed and incorporated into the dUC protocol [15,50]. Although the density of MVs remains unclear, the density of exosomes is 1.13–1.19 g/mL [14]. Upon centrifugation, EVs migrate to the surrounding medium if their densities are same, resulting in further purification of the EVs from other contaminants. For example, the purified exosome pellets from dUC are resuspended into PBS and overlaid on a 30% sucrose cushion with centrifugation at 100,000× *g* [27]. The EV samples can be further fractionated by a step DG using a series of solutions with different density. Iwai et al. used a series of sucrose solutions with concentrations at 2.0, 1.6, 1.18, and 0.8 M and iodixanol solutions with concentration at 50%, 40%, 30%, and 20% to separately isolate exosomes from saliva and collect fractions from different densities [51]. A recent proteomic comparative study was performed to evaluate the dUC and DG and found that DG reduced the presence of co-isolated proteins aggregates and other membranous particles [52]. In comparison to the sucrose gradient, the OptiPrep velocity gradient is reported to perform better at removing some lipoproteins and preserving the size of the vesicles in the gradient [15]. One of the reasons is that the osmotic pressure of sucrose is higher than iodixanol, which could damage EVs in the samples [51].

Some additional strategies are also included in the dUC workflow to increase the purity of EVs for different types of biofluids. THP (also called uromodulin) is a highly abundant protein in urine and can form a polymeric network to trap exosomes during centrifugation at 10,000–20,000× *g*. To alleviate this effect and increase the yield of exosomes, incubation of the crude exosome pellets with dithiothreitol (DTT) or 3-[(3-cholamidopropyl)dimethylammonio]-1-propanesulfonic (CHAPS) were developed. DTT could denature THP, thus inhibiting aggregation and allowing THP to be removed from the supernatant. Moon et al. resuspended the 200,000× *g*-derived urinary pellets in the sucrose solution and incubated with 60 mg/mL DTT at 60 °C for 10 min before DG. A total of 1877 urinary exosome proteins were identified in MS^E analyses [42]. But one of the side effects caused by DTT is that exosomal protein remodeling as DTT is a strong reducing agent and may reduce the exosomal proteins, thus resulting in detrimental effects on their biological activity. Musante et al. used CHAPS which is a mild detergent and known to solubilize THP to replace DTT. They found that CHAPS did not affect vesicle morphology or exosomal marker distribution and preserved better biological activity. Further MS analysis revealed that 76.2% of proteins recovered by CHAPS were identified in those treated by DTT [53]. In addition, Barrachina et al. used KBr in a similar mechanism for plasma samples to reduce lipoproteins in EV samples by solubilizing them [54]. Alternative strategies to improve dUC can be achieved by combinational usage with other types of isolation methods, such as filter device or size exclusion chromatography (SEC). Those combinational methods not only can improve the purity of EVs, but also can dramatically reduce the overall processing time. Details will be presented in the following subsections.

2.3. Size-Based Isolation

Size-based isolation, such as filtration and size exclusion chromatography (SEC), is another type of isolation method, which can be used alone or with other methods to isolate EVs from biofluids. For filtration, samples are passed through a membrane with a specific pore size by centrifugation or pressure. Centrifugation-based filter devices have been reported to yield approximately three-fold greater EVs than that prepared by pressure-driven filter devices [55]. Filters made by different materials have been demonstrated as a fast and simple alternative to dUC. Merchant et al. applied a pore size 0.1 µm of commercially available VVLP (hydrophilized polyvinylidene difluoride) disc membranes to isolate urinary exosomes before MALDI (Matrix-assisted laser desorption/ionization) TOF analysis, and filtration of 50 mL urine samples was achieved within 15 min [56]. Musante et al. developed a "hydrostatic filtration dialysis" process to isolate urinary EVs. Urine samples were centrifuged at 2000× *g* before loaded onto a dialysis membrane with a molecular weight cut-off (MWCO) of 1000 kDa. They found that centrifugation at 2000× *g* allowed to remove the bulk of THP without losing exosomes. By using the dialysis membrane with MWCO of 1000 kDa, solvent, together with all the analytes below 1000 kDa were pushed through the mesh of the membrane due to the hydrostatic pressure of the urine. This method avoided the laborious and time-consuming steps of dUC, while the yield of EVs from this dialysis membrane was reported to outperform the dUC [57,58]. Sequential usage of different types of filters was also explored to isolate EVs. A three-step protocol was established based on sequential steps of dead-end pre-filtration, tangential flow filtration, and low-pressure track-etched membrane filtration. But this sequential filtration step was tested for cell culture, not for biofluids [59]. Instead of used alone, filtration is more commonly used with other types of methods for EV isolation, such as with dUC as a concentration/enrichment step with the aim to concentrate the samples and reduce the processing duration. For example, a 0.22 μm filter device is the most used filter device in EV studies to remove components with a diameter exceeding ca. 200 nm and as one of the processing steps in the dUC [16,60]. In the protocol of Théry et al., the pellets yielded by 2 h of centrifugation at 110,000× g were resuspended in PBS and passed through a 0.22 μm filter before another round of centrifugation at 110,000× g [27]. Shiromizu et al. further simplified the steps by initially using a 300× g centrifugation followed by a filtration step with a 0.22 μm filter to obtain the exosomes crude before a 30% sucrose DG in colorectal cancer biomarker studies [38]. The hydrostatic filtration dialysis can also be used as a pre-enrichment step for dUC to isolate urinary EVs [61].

Despite that the filtration is fast and has the capability of high throughput for EV isolation, the filters can be easily blocked resulted from trapping vesicles or other contaminant aggregates. SEC as another type of size-based isolation strategy has not been normally reported with this limitation posed by filtration [16]. For SEC, samples are loaded onto a column packed with heterogeneous polymeric beads, such as Sepharose, with diverse pore size. In general, the larger molecules are eluted earlier than the smaller ones since the smaller molecules can enter more pores than the larger ones, thus eluted later. Menezes-Neto et al. used SEC as a stand-alone methodology for isolation of EVs. They packed Sepharose CL-2B into a syringe and isolated exosomes from a 1 ml plasma after centrifugation at 500× g for 10 min. A total of 269 proteins were identified from the plasma of one healthy donor on an LTQ Orbitrap Velos mass spectrometer [62]. However, Karimi et al. also packed Sepharose CL-2B beads into a Telos solid phase extraction column and found that this SEC column failed to separate EVs from lipoproteins. Instead of using SEC alone, they overlaid a 6 mL plasma on top of an OptiPrep cushion and centrifuged at 178,000× g before SEC separation. The combinational usage of the density cushion and SEC reduced about 100-fold lipoprotein particles in the EV samples with 1187 proteins identified. [63]. SEC was also reported as an alternative step to replace the final step of dUC. Smolarz et al. used the SEC to isolate exosomes instead of ultracentrifugation at 100,000–200,000× g. Briefly, serum was centrifuged at 1000× g and 10,000× g for 10 and 30 min, respectively. The generated supernatant was filtrated using a 0.22 μm syringe filter unit before loading onto the micro-SEC column to isolate exosomes. A total of 267 proteins were identified by the downstream LC/MS analysis [64]. A commercial size-exclusion chromatography column, qEV, was also used to extract EVs from saliva and tears to study primary Sjögren's syndrome [65]. One of the problems faced by SEC is the increased sample volume obtained after elution, resulting in an extra concentration step for the downstream EV analysis. Foers et al. compared ultracentrifugation and ultrafiltration for the concentration of the SEC eluent. They loaded 10,000× g supernatant of human synovial fluid into a HiPrep 26/60 Sephacryl S-500 HR prepacked gel filtration column. This column contains a hydrophilic, rigid allyl dextran/bisacrylamide matrix and allows for large sample volume input and small EV infiltration. SEC fractions were concentrated by either ultracentrifugation at 100,000× g for 90 min or passing an Amicon Ultra-15 100 kDa cellulose ultrafiltration device. They found ultrafiltration could avoid artifactual aggregation of EVs with contaminants, such as extracellular debris, which were typically observed in samples prepared by ultracentrifugation [66].

2.4. Precipitation-Based Isolation

Polymer precipitation-based isolation has the benefits of commercial availability and easy processing and is now widely applied to isolate EVs from the biofluids under many disease statuses, such as colorectal cancer, epithelial ovarian cancer, and rheumatoid arthritis [67–69]. This type of isolation method is initially used in viral studies by forming a polymer network to decrease the solubility of all components present in the sample [70]. The whole procedure includes mixing an appropriate volume of a polymer solution with samples and incubation. Then, the precipitated EVs are recovered by low-speed centrifugation. The polymer solution could be from a commercial kit, such as ExoQuick, Total Exosome Isolation, and ExoSpin, or home-made polyethylene glycol (PEG) solution [14]. Comparative studies have been conducted to evaluate the EVs isolated by different commercial kits in order to facilitate the choice of isolation methods. Ding et al. compared three commonly used commercial kits for EV isolation, including Total Exosome Isolation, ExoQuick, and RIBO Exosome Isolation Reagent. They found that the size of the majority of particles isolated by those kits was from 30–150 nm, while RIBO generated the highest particle yields. Further western blot (WB) results revealed that ExoQuick was the most efficient method by evaluating the marker proteins of CD63 and TSG101 [71]. Lobb et al. found that ExoSpin performed significantly better in avoiding co-isolation of contaminating proteins and yielded higher levels of EV markers compared to ExoQuick [55].

Although easy–to–use EV commercial kits are now widely used, home-made PEG has relative low-cost of EV preparation. Weng et al. added PEG into samples with a final concentration of 10% and incubated the samples at 4 °C for 2 h before recovery at centrifugation of 3000× *g*. Then a second-round of PEG precipitation was followed in order to improve the purity of EVs. The downstream MS analysis identified a total of 6299 protein groups from HeLa cell culture supernatant. Unfortunately, they did not test any biofluid sample in the study [72]. PEG has also been demonstrated to be used together with ultracentrifugation. Rider et al. purified the EVs resulted from one-round of PEG precipitation by further centrifugation at 100,000× *g* for 70 min [73]. Instead of isolating EVs by precipitation, aqueous two-phase systems (ATPSs) were proposed by Shin et al. They used a PEG/dextran ATPS to isolate EVs from the tumor interstitial fluid based on the mechanism that different kinds of particles are effectively partitioned to different phases in a short time. Their comparative studies showed that ATPSs could recovery about 70% of EVs from the EV protein mixtures, whereas the recovery for dUC and ExoQuick were about 16% and 40% [74]. But one should notice that EVs isolated by precipitation may be contaminated by polymer molecules, such as PEG, which is well-known for interfering in MS-based proteomic analysis. Therefore, it is necessary to remove those polymer molecules before MS analysis.

2.5. *Affinity-Based Isolation*

Apart from size and density, EVs share some common characteristics, like general protein composition and lipid bilayer structure. By utilizing those common characteristics, affinity-based isolation could achieve the isolation of EVs from complex biological samples. The main principle of affinity-based isolation is via the interaction between the surface markers of EVs with the antibody, molecules, or function group immobilized onto various carriers to separate EVs from the analyzed biofluids. Among those methods, immuno-based isolation is the most widely available and used method [15,75]. Some proteins have often been used as exosome-associated markers including the tetraspanin family (such as CD8, CD9, CD61, CD63, CD81, and CD82), cytoplasmic proteins (such as tubulin, actin, actin-binding proteins, annexins, and Rab proteins), and heat shock proteins (such as Hsp70, and Hsp90). Therefore, the antibodies against those common proteins coupled to different carriers have been utilized to isolate EVs [76–78]. Hildonen at el. isolated urinary exosomes from healthy subjects by immunocapture on magnetic beads. They coupled the antibody cocktail against CD8, CD61, and CD81 to magnetic beads. By digestions on beads in non-detergent containing buffer, they studied the outer membrane-associated proteins of exosomes and found 49 proteins associated or bound to membranes [76]. Antibody against tetraspanins was also shown to immobilize on highly porous monolithic silica microtips and applied to investigate lung cancer biomarker proteins on exosomes in serum samples. The subsequent MS analysis had identified 1369 proteins [77]. In addition to those common markers of EVs, immuno-based isolation was also explored to isolate the desired groups of EVs because the function of EVs appears to be determined by its specific protein content. For example, anti-EpCAM-coupled microbeads were employed to extract epithelial tumor-derived

EVs from plasma since it has been demonstrated that exosomes from epithelial tumors express EpCAM (epithelial cell adhesion molecule) on their surface [78,79]. Tauro et al. isolated two distinct populations of exosomes released from organoids derived from the human colon carcinoma cell line LIM1863EVs, via sequential immunocapture using anti-A33- and anti-EpCAM-coupled magnetic beads [80].

In addition to antibodies, some EV-binding molecules, such as specific peptides including venceremin or Vn, and heparin, were also investigated to isolate EVs [14]. Vn, a novel class of peptides, which exhibit the specific affinity for heat shock proteins were selected for isolation of EVs from breast cancer [81]. Bijnsdorp et al. compared the urinary EVs isolated by Vn-96 and dUC and found that more than 85% of the proteins were identified both in EVs isolated by Vn and dUC. But the Vn96-peptide offered easier and time convenient methods in comparison with dUC [82]. Heparin is a highly sulfated glycosaminoglycan and has recently been used to isolate the EVs in which the surface contains the cell surface receptor, heparan sulfate proteoglycans. Balaj et al. incubated plasma with heparin-coated beads overnight and further processed the enriched samples by ultracentrifuging at 100,000× g for 90 min or a 100 kDa MWCO filter. The EVs isolated by heparin-affinity beads were detected to contain the EV marker of Alix and lower level of protein contamination [83].

Affinity for targeted proteins on the surface of EVs can be problematic for general EV studies since an unreliable analysis could be obtained due to the exclusion of EVs without targeted proteins.

Therefore, an affinity for the lipid membrane structures of EVs is utilized. Gao et al. recently adopted the TiO_2 material, which is commonly used for the enrichment of phosphopeptides to isolate EVs. Through the interaction with the phosphate groups on the lipid bilayer of EVs, TiO_2 can enrich EVs from serum within 5 min [84]. Tan et al. also focused on the membrane lipid as the target and used phospholipid-binding ligands to extract plasma EVs. Based on previous studies, EVs could be differentiated by their membrane phospholipid composition, specifically GM1 gangliosides and phosphatidylserines. They found two distinct groups of EVs by using cholera toxin B chain (CTB) and annexin V (AV), which, respectively, binds GM1 ganglioside and phosphatidylserine [85]. Nakai et al. developed a novel method for EV purification by using Tim4 proteins. Tim4 proteins can capture EVs via the specific interaction with the phosphatidylserine displayed on the surface of EVs and release the EVs by adding Ca^{2+} chelators. They claimed that the lower contaminations were found in the EV samples isolated by Tim4 proteins [86].

3. Comparative Studies for Isolation Methods of EVs

Among the isolation methods discussed above, it is generally thought that dUC is time-consuming. Filtration has the risk of stuck EVs in the membrane pores, while SEC is not ideal for large scale isolation. Although precipitation-based and immuno-based methods usually involve easy processing, the purity of EVs from precipitation is often problematic and affinity-based isolation is often considered as a good technique for isolation of sub-populations of EVs [16]. However, it is more reasonable to evaluate each isolation method based on the detailed protocol used and criteria of evaluation in each study. Otherwise, purity, efficiency, and reproducibility of different isolations could easily confound literature. For example, Kalra et al. performed a comparative evaluation of three exosome isolation techniques: dUC, anti-EpCAM conjugated microbeads, and OptiPrep DG. Their results suggested that the OptiPrep DG was superior in isolating pure exosomal populations by comparing the level of highly abundant plasma proteins which were detected by MS in the isolated plasma EV samples [79]. Those three methods were also compared by Greening et al. in a cell model. Based on the quantitative MS results for the identified exosome markers and proteins associated with EV biogenesis, trafficking, and release, anti-EpCAM was shown to be the most effective method to isolate exosomes [50]. Results from those two comparative studies can be explained by the differences in the sample types, details of protocols, and criteria of evaluation used in each study. Therefore, the selected studies for evaluation of different EV isolation methods are listed in Table 2 for better interpretation of each isolation. One thing to be mentioned is that the comparative studies listed in Table 2 also include the studies based on cell cultures, animals, and characterization of EVs by other methods, and are not just based on biofluid samples and analyses of MS.

Table 2. Selected comparative studies for EV isolation.

Isolation Methods	Characterization Techniques	Samples	Study
dUC, SEC	NTA, Dissociation-enhanced lanthanide fluorescence immunoassay, WB, TEM	rat plasma, cell culture	[87]
dUC, SEC	TEM, AFM, WB, MS	cell culture	[88]
Affinity-based (exoEasy kit) and SEC (qEV)	WB, TEM, NTA, lipid quantification kit, RNA quality	plasma	[89]
dUC and Commercial kit from Invitrogen, 101Bio, Wako and iZON	Dynamic Light Scattering, immunoblot analysis, qRT-PCR, MS, Cell Proliferation Assay	cell culture	[90]
dUC, precipitation (ExoQuick, Total Exosome Isolation Reafent, Exo-PREP) and SEC (qEV)	TEM, NTA, WB	cell culture	[91]
Lectin-based, Exoquick, Total exosome Isolation and in-house modified procedure	WB, Reverse transcriptase and qPCR, EM	urine	[92]
dUC, precipitation (ExoQuick, Total exosome isolation, PEG, Exo-spin), filtration (ExoMir)	NTA, Flow cytometry, WB, PCR,	serum	[93]
dUC, filtration (Stirred cell and Centricon), OptiPrep DG, ExoQuick, Exo-spin, SEC	Tunable resistive pulse sensing, EM, WB	cell culture and plasma	[55]
SEC and Exo-Spin	NTA, Flow cytometry, MS	plasma	[62]
dUC, anti-EpCAM, OptiPrep DG	MS, WB, TEM	plasma	[79]
Nanomembrane ultrafiltration, dUC and dUC-SEC	MS, TEM, WB	urine	[94]
dUC, anti-EpCAM, OptiPrep DG	TEM, CryoEM, MS	cell culture	[50]
Sucrose DG and ExoQuick	TEM, NTA, WB	serum	[95]

* EM: electron microscopy; TEM: transmission electron microscopy; NTA: nanoparticle tracking analysis AFM: atomic force microscopy; WB: western blot.

As shown in Table 2, many studies have compared the EV isolation by different techniques; thus, according to different criteria. Different criteria were also applied, even if the same technique was used for assessment [55,88,92,94]. WB for EV marker proteins is one of the commonly used methods to compare the efficiency of EV isolation. But how many and which marker proteins should be chosen for the good comparison has not been well established. Lobb et al. provided a comparative analysis of four EV isolation techniques. dUC, ultrafiltration, SEC, OptiPrep DG, and precipitation (ExoQuick and ExoSpin) were used to isolate EVs from cell culture and plasma. By comparing the levels of exosomal markers of HSP70, Flotillin-1, and TSG 101 in WB, precipitation protocols provided the least pure preparations of EVs, whereas SEC isolation was comparable to DG purification of EVs [55]. In a similar way, Royo et al. tested the EV isolation of lectin-based purification, Exoquick, Total Exosome Isolation, and an in-house modified EV isolation procedure via WB of eight EV protein markers including CD9, CD10, CD63, TSG101, CD10, AIP1/Alix, AQP2, and FLT1. They observed that the levels of different EV marker proteins varied by different isolations and, thus, suggested that different methods isolated a different mixture of urinary EV marker proteins [92]. Evaluation of EV isolation by MS also lacks criteria to make a universal, comprehensive comparison. Rood et al. centrifugated the urine samples at 17,000× g for 15 min and then isolated the EVs by further centrifuging at 200,000× g for 110 min or filtering with 100 kDa Vivaspin 20 polyethersulfone nanomembrane concentrators. They found that either ultracentrifugation or ultrafiltration was difficult to isolate EVs from urine since highly abundant proteins, especially albumin and α-1-antitrypsin, were present in large amounts, which significantly limited the detection of MALDI-TOF. Additional SEC following ultracentrifugation was suggested to use in order to improve the purity of EVs [94]. Based on the gene ontology analysis for the identified proteins by MS, Davis et al. believed that dUC and SEC did not isolate equivalent EV population profiles [88]. Altogether, cautions should be taken when interpreting each EV isolation.

Rather than focus on the performance in yield or purity of each isolation, the functional activity of EVs was also reported to depend on the isolation method used [87,91]. Antounians et al. noticed that amniotic fluid stem cell-derived EVs isolated by dUC, precipitation (ExoQuick, Total Exosome Isolation Reagent, and Exo-PREP), and SEC (qEV column) had different effects on a model of damaged lung epithelium [91]. It suggests the necessity of evaluating the isolation methods within the content of biology.

4. MS Strategies Used in Proteomic Studies of Extracellular Vesicles

4.1. Sample Preparation and Separation

To date, proteomic studies of EVs are mainly conducted based on the bottom-up MS strategy. As shown in Figure 2, protein should be extracted from the isolated EVs and digested before MS analysis. For proteomic analysis, EV proteins are commonly extracted using the lysis buffer with detergent (such as sodium dodecyl sulfate (SDS)) or without detergent (such as 8 M urea). TRIzol reagent, which is often used in isolation of nucleic acid from EVs, has been recently reported to extract proteins from EVs. Joy et al. compared the EV protein extraction between Laemmli and TRIzol. Laemmli buffer typically contains 2% SDS, 10% glycerol in Tris-HCl with pH 6.8, which is an effective protein-extraction for EVs. They found that these two methods gave similar results in their ability to extract proteins and ~60% of proteins were identified in the samples prepared by both methods. However, they did not apply TRIzol reagent on any EV samples from biofluids [96]. Special extraction methods are also investigated to facilitate studies of sub-populations of proteins in the EVs, such as membrane proteins. Hu et al. optimized the Triton X-114 detergent partitioning protocol to target the analysis of membrane proteins of urinary EVs. Dried EV pellets were dissolved in 1% SDS containing lysis buffer for 1 h before adding 2.2% pre-condensed Triton X-114 buffer. A lower detergent phase, with an oily appearance, and an upper aqueous phase were formed when the temperature was above the clouding point of Triton X-114. Proteins in each phase were recovered by acetone precipitation before MS analysis. Most of the membrane proteins of urinary EVs were found in the detergent fraction [58].

As shown in Table 1, filter aided sample preparation (FASP) was utilized in some EV studies to achieve an easy process for buffer exchange and protein digestion [97]. In FASP, the extracted EV proteins are transferred into a molecular weight cut-off filter. This filter can retain most of the proteins on the membrane after simple centrifugation. Meanwhile, peptides can freely pass through the membrane during centrifugation. By using this kind of filter, the denaturing detergent-based buffer used for protein extraction can be easily changed to a digestion buffer, and the sample can be digested on the filter without extra transferring steps. FASP, with easy processing and minimal sample loss, has become the method of choice in many EV studies, especially in the limited amount of samples available [16]. Fel et al. improved the FASP by using multi-enzyme digestion to prepare EV samples obtained by precipitation. In their studies, serum samples from polycythemia vera patients were centrifuged at 2000× *g* for 30 min to remove cells and debris before incubation with the reagent from the Total Exosome Isolation kit. Afterward, the proteins were extracted from EVs and digested sequentially by Lys C, trypsin, and chymotrypsin in a Micron 30 kDa filter (Figure 4). A total of 706 proteins were identified with thirty-eight proteins showing significant differences in the patients' group [97].

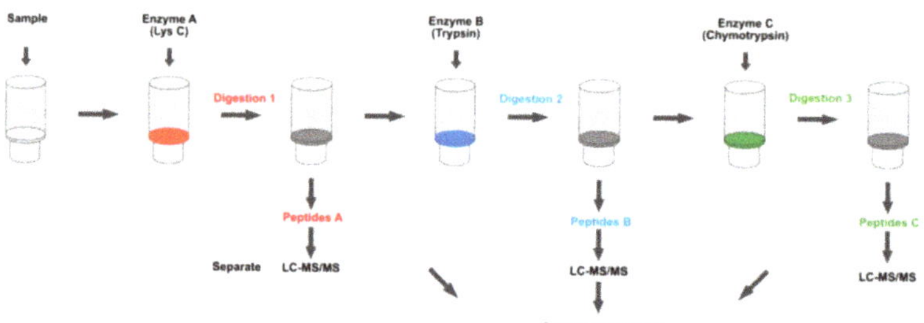

Figure 4. The schematic workflow for multi-enzyme digestion filter-aided sample preparation. This figure was adopted from Ref. [97].

To perform in-depth proteomic analysis, additional separation before LC/MS analysis can be performed by either gel electrophoresis or liquid chromatography. Gel electrophoresis can effectively remove the most common contaminants in the samples according to the molecular weight of proteins, which could benefit the downstream MS analysis. Both Tsuno et al. and Xie et al. isolated EVs from serum using ExoQuick and separated the protein content through two-dimensional gel electrophoresis before MALDI-TOF analysis to study rheumatoid arthritis and coronary artery aneurysms, respectively [69,98]. Gel electrophoresis has also been applied to study EVs from urine, breast milk, and saliva [45,47,99]. Apart from separation based on gel electrophoresis, two-dimensional liquid chromatography (2D-LC) is utilized to analyze EV samples [30,38–40,100]. Antwi-Baffour et al. isolated MVs from the plasma of malaria patients and used a microcapillary strong cation exchange (SCX) column to fractionate the digested MVs samples. A total of 1729 proteins were identified in malaria samples, while only 234 proteins were identified in healthy control samples [30]. Their finding may imply that MVs in disease status could result in more protein identification than in healthy. Shiromizu et al. further simplified the fractionation of EV samples by using a C18-SCX Stage-tip. Using this strategy, they identified 702 proteins from the serum of colorectal cancer patients [38]. Instead of SCX as the first-dimensional separation, Lin et al. performed a high pH reverse phase chromatography to fractionate EVs from semen and study asthenozoospermia with 3699 protein identified by MS [40].

In addition to the typical proteomic studies, separation methods vary according to different studies, such as the studying of post-translational modifications of EV proteins. The electrostatic repulsion-hydrophilic interaction chromatography (ERLIC) was employed to facilitate the study of glycoproteins from EVs. Cheow et al. centrifuged plasma at 100,000× g for 2 h and 200,000× g for 18 h. They recovered a visible yellow suspension that was highly enriched in soluble glycoproteins and EVs. After protein extraction and digestion, an ERLIC column was used to simultaneously enrich secretory and EV-enriched glycoproteins and further fractionate the sample. A total of 127 plasma glycoproteins were identified with high confidence [101]. In order to study N-linked glycoproteomics of urinary exosomes, Saraswat et al. isolated urinary EVs by centrifugation at 200,000× g for 2 h and applied SNA affinity chromatography or SEC to enrich glycopeptides in the urinary EVs after tryptic digestion. In total, 126 *N*-glycopeptides from 51 *N*-glycosylation sites belonging to 37 glycoproteins were found [102].

4.2. MS Acquisition

During MS analysis, data-dependent acquisition (DDA) are normally used. Recently, data-independent acquisitions (DIA), such as SWATH (sequential window acquisition of all theoretical fragment ion), MS^E, and multiplexed MS/MS, are used in EV studies to satisfy different purposes. Unlike DDA, DIA simultaneously fragments all precursor ions present in a wide isolation window. Braga-Lagache et al. analyzed MV proteins from plasma samples by both

DDA and multiplexed DIA on a quadrupole orbitrap instrument. In each cycle of multiplexed DIA, data is usually acquired with one full MS scan followed by a series of MS2, such as ten MS2 scans. Each MS2 scan records all the fragment ions generated by precursor ions that are isolated from multiple different isolation windows with a fixed *m/z* range, such as isolated from three randomly combined 10 *m/z* isolation windows. A targeted approach is used to analyze the DIA data by using spectral libraries from formerly acquired fragment spectra with exact mass and retention time of precursors. They found that a multiplexed DIA approach only consumed one third of the DDA acquisition time when data was extracted by a targeted approach. Their results suggested that multiplexed DIA was a valuable alternative to DDA [103]. Moon et al. and Chutipongtanate et al. also applied DIA to analyze the protein content of EVs [32,42]. In the study of Moon et al., crude exosomes prepared by sucrose density ultracentrifugation were digested in-gel and analyzed by MS^E on a Waters Q-TOF mass spectrometer. In MS^E, alternating low- and high-energy collision-induced dissociation are used. The low-energy scan is used to obtain precursor information, while the high-energy scan is to collect fragment ions. A total of 1877 urinary exosome proteins were identified from IgA nephropathy and thin basement membrane nephropathy patients [42]. Chutipongtanate et al. utilized SWATH to analyze urinary EV proteins. In SWATH, the mass range of interest is divided into several segments with a fixed *m/z* range, such as 25 *m/z*. Then, precursor ions within each segment are fragmented together until all the segments are analyzed. They achieve a label-free DIA quantitative analysis for EV and MV proteins with a curated spectral library of 1145 targets, suggesting their potential clinical use [32].

Quantitative MS based on label and label-free have been demonstrated to study various diseases, such as prostate cancer, asthenozoospermia and venous thrombosis [39,40,46,104]. Fujita et al. labeled the urinary EV proteins with isobaric tag for relative and absolute quantitation (iTRAQ). A total of 4710 proteins were identified by MS, including 3528 proteins quantified [39]. Lin et al. quantified seminal EV proteins with iTRAQ labeling and revealed 91 proteins with significant changes [40]. 2D-LC and tandem mass tag (TMT) were also used to quantitative analysis of EVs in HIV-infected alcohol drinkers and cigarette smokers through precipitation-based isolation [104]. Although stable isotope labeling by amino acids in cell culture (SILAC) cannot label EV proteins from human biofluids, a PROMIS-Quan method which based on SILAC quantification was developed in order to gain a comprehensive quantification for potential clinical EV protein analysis. In PROMIS-Quan, EV lysates were spiked with super-SILAC which was prepared from cell cultures and served as an internal standard. Then, the same set of super-SILAC mix was quantified relative to purified proteins of interest, with known absolute amounts. By this way, EV proteins can be quantified not only in large-scale but also retrospectively only relative to the same set of super-SILAC standard [29]. Quantitative MS is not only applied to the EV studies with the aim of biomarker discovery but also developed as an evaluation method to assess the EV isolation. Wang et al. established a multiple reaction monitoring (MRM) based method to assess the purity of EVs. MRM is often used for target quantitative analysis as a validation method for biomarkers reported in discovery MS analysis. They first generated ^{15}N-labeled quantification concatamers (QconCATs) for a pattern of targeted EV proteins and abundant serum proteins (non-EV proteins or contaminants) as the internal standards for quantification of those proteins in MRM. QconCATs were artificial proteins composed of concatenated tryptic peptides from targeted proteins. The purity of EVs was then assessed by the quantitative results of the targeted EV proteins and abundant serum proteins in MRM [105]. They further expanded this method to separate EVs and lipoprotein particles by adding QconCAT for apolipoproteins into the previous MRM assay [106].

5. Conclusions

With a greater understanding of the roles of EVs in the regulation of physiological and pathological processes, an increased need to use that knowledge for diagnosis and therapy of diseases has emerged. To satisfy that increased need, establishing an EV isolation method that provides rapid, efficient, and high throughput isolation and enables assessment of the full spectrum of EVs is required. Unfortunately, the currently available isolation methods only partially meet the requirement. MS is a powerful tool for the characterization of the protein content of EVs, which is crucial to decipher the biological role of EVs and explore their potential use as diagnostic, monitoring, and therapeutic tools. Currently, the application of MS in EV studies is largely limited by the imperfections of EV isolation methods.

The increasing number of studies have pointed out the EV samples prepared by current isolation methods containing different sub-populations of EVs and contaminants from surroundings. Contaminants in the isolated EV samples may not only cover the signal of lower abundant EV proteins during MS analysis but also increase the difficulty of MS data analysis, since there is no current standard to clearly distinguish EV proteins from contaminants, especially the uncommon contaminants, in the MS-generated list. To address those problems, future improvements on EV isolation and MS analysis are urgently required.

Author Contributions: Conceptualization: C.Y.; writing and original draft preparation: C.Y., J.L., X.H., and Y.D.; review and editing: C.Y. and J.L.

Funding: This research was funded by the National Natural Science Foundation of China (21703163) and the Fundamental Research Funds for the Central Universities (2018IVB043A).

Conflicts of Interest: The authors declare no conflict of interest.

References

1. Merchant, M.L.; Rood, I.M.; Deegens, J.K.J.; Klein, J.B. Isolation and Characterization of Urinary Extracellular Vesicles: Implications for Biomarker Discovery. *Nat. Rev. Nephrol.* **2017**, *13*, 731–749. [CrossRef] [PubMed]
2. Wolf, P. The Nature and Significance of Platelet Products in Human Plasma. *Br. J. Haematol.* **1967**, *13*, 269–288. [CrossRef] [PubMed]
3. Yuana, Y.; Sturk, A.; Nieuwland, R. Extracellular Vesicles in Physiological and Pathological Conditions. *Blood Rev.* **2013**, *27*, 31–39. [CrossRef] [PubMed]
4. Quinn, J.F.; Patel, T.; Wong, D.; Das, S.; Freedman, J.E.; Laurent, L.C.; Carter, B.S.; Hochberg, F.; Van Keuren-Jensen, K.; Huentelman, M.; et al. Extracellular Rnas: Development as Biomarkers of Human Disease. *J. Extracell Vesicles* **2015**, *4*, 27495. [CrossRef] [PubMed]
5. Loyer, X.; Vion, A.C.; Tedgui, A.; Boulanger, C.M. Microvesicles as Cell-Cell Messengers in Cardiovascular Diseases. *Circ. Res.* **2014**, *114*, 345–353. [CrossRef] [PubMed]
6. Pocsfalvi, G.; Stanly, C.; Vilasi, A.; Fiume, I.; Capasso, G.; Turiák, L.; Buzas, E.I.; Vékey, K. Mass Spectrometry of Extracellular Vesicles. *Mass Spectrom. Rev.* **2016**, *35*, 3–21. [CrossRef] [PubMed]
7. Simonsen, J.B. What Are We Looking At? Extracellular Vesicles, Lipoproteins, or Both. *Circ. Res.* **2017**, *121*, 920–922. [CrossRef] [PubMed]
8. Simpson, R.J.; Kalra, H.; Mathivanan, S. Exocarta as a Resource for Exosomal Research. *J. Extracell Vesicles* **2012**, *1*, 18374. [CrossRef] [PubMed]
9. Kalra, H.; Simpson, R.J.; Ji, H.; Aikawa, E.; Altevogt, P.; Askenase, P.; Bond, V.C.; Borras, F.E.; Breakefield, X.; Budnik, V.; et al. Vesiclepedia: A Compendium for Extracellular Vesicles with Continuous Community Annotation. *PLoS Biol.* **2012**, *10*, e1001450. [CrossRef] [PubMed]
10. Kim, D.K.; Lee, J.; Kim, S.R.; Choi, D.S.; Yoon, Y.J.; Kim, J.H.; Go, G.; Nhung, D.; Hong, K.; Jang, S.C.; et al. Evpedia: A Community Web Portal for Extracellular Vesicles Research. *Bioinformatics* **2015**, *31*, 933–939. [CrossRef] [PubMed]
11. Witwer, K.W.; Buzas, E.; Bemis, L.T.; Bora, A.; Lasser, C.; Lotvall, J.; Nolte-t Hoen, E.N.; Piper, M.G.; Sivaraman, S.; Skog, J.; et al. Standardization of Sample Collection, Isolation and Analysis Methods in Extracellular Vesicle Research. *J. Extracell Vesicles* **2013**, *2*, 20360. [CrossRef] [PubMed]
12. Gould, S.J.; Raposo, G. As We Wait: Coping with an Imperfect Nomenclature for Extracellular Vesicles. *J. Extracell Vesicles* **2013**, *2*, 20389. [CrossRef] [PubMed]
13. Barrachina, M.N.; Calderon-Cruz, B.; Fernandez-Rocca, L.; Garcia, A. Application of Extracellular Vesicles Proteomics to Cardiovascular Disease: Guidelines, Data Analysis, and Future Perspectives. *Proteomics* **2019**, *19*, 1800247. [CrossRef] [PubMed]
14. Wang, W.; Luo, J.; Wang, S. Recent Progress in Isolation and Detection of Extracellular Vesicles for Cancer Diagnostics. *Adv. Healthc. Mater.* **2018**, *7*, e1800484. [CrossRef] [PubMed]

15. Szatanek, R.; Baran, J.; Siedlar, M.; Baj-Krzyworzeka, M. Isolation of Extracellular Vesicles: Determining the Correct Approach. *Int. J. Mol. Med.* **2015**, *36*, 11–17. [CrossRef] [PubMed]
16. Abramowicz, A.; Widlak, P.; Pietrowska, M. Proteomic Analysis of Exosomal Cargo: The Challenge of High Purity Vesicle Isolation. *Mol. Biosyst.* **2016**, *12*, 1407–1419. [CrossRef] [PubMed]
17. Tzouanas, C.; Lim, J.S.Y.; Wen, Y.; Thiery, J.P.; Khoo, B.L. Microdevices for Non-Invasive Detection of Bladder Cancer. *Chemosensors* **2017**, *5*, 30. [CrossRef]
18. Yuana, Y.; Boing, A.N.; Grootemaat, A.E.; van der Pol, E.; Hau, C.M.; Cizmar, P.; Buhr, E.; Sturk, A.; Nieuwland, R. Handling and Storage of Human Body Fluids for Analysis of Extracellular Vesicles. *J. Extracell. Vesicles* **2015**, *4*, 29260. [CrossRef] [PubMed]
19. Lacroix, R.; Judicone, C.; Mooberry, M.; Boucekine, M.; Key, N.S.; Dignat-George, F.; The ISTH SSC Workshop. Standardization of Pre-Analytical Variables in Plasma Microparticle Determination: Results of the International Society on Thrombosis and Haemostasis Ssc Collaborative Workshop. *J. Thromb. Haemost.* **2013**, *11*, 1190–1193. [CrossRef] [PubMed]
20. Yuana, Y.; Bertina, R.M.; Osanto, S. Pre-Analytical and Analytical Issues in the Analysis of Blood Microparticles. *Thromb. Haemost.* **2011**, *105*, 396–408. [CrossRef] [PubMed]
21. Akers, J.C.; Ramakrishnan, V.; Yang, I.; Hua, W.; Mao, Y.; Carter, B.S.; Chen, C.C. Optimizing Preservation of Extracellular Vesicular Mirnas Derived from Clinical Cerebrospinal Fluid. *Cancer Biomark.* **2016**, *17*, 125–132. [CrossRef] [PubMed]
22. Jamaly, S.; Ramberg, C.; Olsen, R.; Latysheva, N.; Webster, P.; Sovershaev, T.; Braekkan, S.K.; Hansen, J.B. Impact of Preanalytical Conditions on Plasma Concentration and Size Distribution of Extracellular Vesicles Using Nanoparticle Tracking Analysis. *Sci. Rep.* **2018**, *8*, 17216. [CrossRef] [PubMed]
23. Jeyaram, A.; Jay, S.M. Preservation and Storage Stability of Extracellular Vesicles for Therapeutic Applications. *AAPS J.* **2017**, *20*, 1. [CrossRef] [PubMed]
24. Ge, Q.Y.; Zhou, Y.X.; Lu, J.F.; Bai, Y.F.; Xie, X.Y.; Lu, Z.H. Mirna in Plasma Exosome Is Stable under Different Storage Conditions. *Molecules* **2014**, *19*, 1568–1575. [CrossRef] [PubMed]
25. Gardiner, C.; Di Vizio, D.; Sahoo, S.; Thery, C.; Witwer, K.W.; Wauben, M.; Hill, A.F. Techniques Used for the Isolation and Characterization of Extracellular Vesicles: Results of a Worldwide Survey. *J. Extracell. Vesicles* **2016**, *5*, 32945. [CrossRef] [PubMed]
26. Livshts, M.A.; Khomyakova, E.; Evtushenko, E.G.; Lazarev, V.N.; Kulemin, N.A.; Semina, S.E.; Generozov, E.V.; Govorun, V.M. Isolation of Exosomes by Differential Centrifugation: Theoretical Analysis of a Commonly Used Protocol. *Sci Rep.* **2015**, *5*, 17319. [CrossRef] [PubMed]
27. Thery, C.; Amigorena, S.; Raposo, G.; Clayton, A. Isolation and Characterization of Exosomes from Cell Culture Supernatants and Biological Fluids. *Curr. Protoc. Cell Biol.* **2006**, *30*, 3–22. [CrossRef] [PubMed]
28. Momen-Heravi, F.; Balaj, L.; Alian, S.; Trachtenberg, A.J.; Hochberg, F.H.; Skog, J.; Kuo, W.P. Impact of Biofluid Viscosity on Size and Sedimentation Efficiency of the Isolated Microvesicles. *Front. Physiol.* **2012**, *3*, 162. [CrossRef]
29. Harel, M.; Oren-Giladi, P.; Kaidar-Person, O.; Shaked, Y.; Geiger, T. Proteomics of Microparticles with Silac Quantification (Promis-Quan): A Novel Proteomic Method for Plasma Biomarker Quantification. *Mol. Cell Proteom.* **2015**, *14*, 1127–1136. [CrossRef]
30. Antwi-Baffour, S.; Adjei, J.K.; Agyemang-Yeboah, F.; Annani-Akollor, M.; Kyeremeh, R.; Asare, G.A.; Gyan, B. Proteomic Analysis of Microparticles Isolated from Malaria Positive Blood Samples. *Proteome Sci.* **2017**, *15*, 5. [CrossRef] [PubMed]
31. Raposo, G.; Nijman, H.W.; Stoorvogel, W.; Liejendekker, R.; Harding, C.V.; Melief, C.J.; Geuze, H.J. B Lymphocytes Secrete Antigen-Presenting Vesicles. *J. Exp. Med.* **1996**, *183*, 1161–1172. [CrossRef] [PubMed]
32. Chutipongtanate, S.; Greis, K.D. Multiplex Biomarker Screening Assay for Urinary Extracellular Vesicles Study: A Targeted Labelfree Proteomic Approach. *Sci. Rep.* **2018**, *8*, 15039. [CrossRef] [PubMed]
33. Sun, Y.; Huo, C.; Qiao, Z.; Shang, Z.; Uzzaman, A.; Liu, S.; Jiang, X.; Fan, L.; Ji, L.; Guan, X.; et al. Comparative Proteomic Analysis of Exosomes and Microvesicles in Human Saliva for Lung Cancer. *J. Proteome Res.* **2018**, *17*, 1101–1107. [CrossRef] [PubMed]
34. Whitham, M.; Parker, B.L.; Friedrichsen, M.; Hingst, J.R.; Hjorth, M.; Hughes, W.E.; Egan, C.L.; Cron, L.; Watt, K.I.; Kuchel, R.P.; et al. Extracellular Vesicles Provide a Means for Tissue Crosstalk During Exercise. *Cell Metab.* **2018**, *27*, 237–251. [CrossRef] [PubMed]

35. Kim, J.; Tan, Z.; Lubman, D.M. Exosome Enrichment of Human Serum Using Multiple Cycles of Centrifugation. *Electrophoresis* **2015**, *36*, 2017–2026. [CrossRef] [PubMed]
36. Langevin, S.M.; Kuhnell, D.; Orr-Asman, M.A.; Biesiada, J.; Zhang, X.; Medvedovic, M.; Thomas, H.E. Balancing Yield, Purity and Practicality: A Modified Differential Ultracentrifugation Protocol for Efficient Isolation of Small Extracellular Vesicles from Human Serum. *RNA Biol.* **2019**, *16*, 5–12. [CrossRef] [PubMed]
37. Cvjetkovic, A.; Lotvall, J.; Lasser, C. The Influence of Rotor Type and Centrifugation Time on the Yield and Purity of Extracellular Vesicles. *J. Extracell. Vesicles* **2014**, *3*, 23111. [CrossRef] [PubMed]
38. Shiromizu, T.; Kume, H.; Ishida, M.; Adachi, J.; Kano, M.; Matsubara, H.; Tomonaga, T. Quantitation of Putative Colorectal Cancer Biomarker Candidates in Serum Extracellular Vesicles by Targeted Proteomics. *Sci. Rep.* **2017**, *7*, 12782. [CrossRef] [PubMed]
39. Fujita, K.; Kume, H.; Matsuzaki, K.; Kawashima, A.; Ujike, T.; Nagahara, A.; Uemura, M.; Miyagawa, Y.; Tomonaga, T.; Nonomura, N. Proteomic Analysis of Urinary Extracellular Vesicles from High Gleason Score Prostate Cancer. *Sci. Rep.* **2017**, *7*, 42961. [CrossRef]
40. Lin, Y.; Liang, A.; He, Y.; Li, Z.; Li, Z.; Wang, G.; Sun, F. Proteomic Analysis of Seminal Extracellular Vesicle Proteins Involved in Asthenozoospermia by iTRAQ. *Mol. Reprod Dev.* **2019**. [CrossRef]
41. Jiao, Y.J.; Jin, D.D.; Jiang, F.; Liu, J.X.; Qu, L.S.; Ni, W.K.; Liu, Z.X.; Lu, C.H.; Ni, R.Z.; Zhu, J.; et al. Characterization and Proteomic Profiling of Pancreatic Cancer-Derived Serum Exosomes. *J. Cell Biochem.* **2019**, *120*, 988–999. [CrossRef] [PubMed]
42. Moon, P.G.; Lee, J.E.; You, S.; Kim, T.K.; Cho, J.H.; Kim, I.S.; Kwon, T.H.; Kim, C.D.; Park, S.H.; Hwang, D.; et al. Proteomic Analysis of Urinary Exosomes from Patients of Early Iga Nephropathy and Thin Basement Membrane Nephropathy. *Proteomics* **2011**, *11*, 2459–2475. [CrossRef] [PubMed]
43. Jia, R.; Li, J.; Rui, C.; Ji, H.; Ding, H.; Lu, Y.; Dc, W.; Sun, L. Comparative Proteomic Profile of the Human Umbilical Cord Blood Exosomes between Normal and Preeclampsia Pregnancies with High-Resolution Mass Spectrometry. *Cell Physiol. Biochem.* **2015**, *36*, 2299–2306. [CrossRef] [PubMed]
44. Pisitkun, T.; Gandolfo, M.T.; Das, S.; Knepper, M.A.; Bagnasco, S.M. Application of Systems Biology Principles to Protein Biomarker Discovery: Urinary Exosomal Proteome in Renal Transplantation. *Proteomics Clin. Appl.* **2012**, *6*, 268–278. [CrossRef] [PubMed]
45. Winck, F.V.; Ribeiro, A.C.P.; Domingues, R.R.; Ling, L.Y.; Riano-Pachon, D.M.; Rivera, C.; Brandao, T.B.; Gouvea, A.F.; Santos-Silva, A.R.; Coletta, R.D.; et al. Insights into Immune Responses in Oral Cancer through Proteomic Analysis of Saliva and Salivary Extracellular Vesicles. *Sci. Rep.* **2015**, *5*, 16305. [CrossRef] [PubMed]
46. Ramacciotti, E.; Hawley, A.E.; Wrobleski, S.K.; Myers, D.D., Jr.; Strahler, J.R.; Andrews, P.C.; Guire, K.E.; Henke, P.K.; Wakefield, T.W. Proteomics of Microparticles after Deep Venous Thrombosis. *Thromb. Res.* **2010**, *125*, e269–e274. [CrossRef] [PubMed]
47. Van Herwijnen, M.J.; Zonneveld, M.I.; Goerdayal, S.; Nolte-'t Hoen, E.N.; Garssen, J.; Stahl, B.; Maarten Altelaar, A.F.; Redegeld, F.A.; Wauben, M.H. Comprehensive Proteomic Analysis of Human Milk-Derived Extracellular Vesicles Unveils a Novel Functional Proteome Distinct from Other Milk Components. *Mol. Cell. Proteom.* **2016**, *15*, 3412–3423. [CrossRef]
48. Chen, I.H.; Xue, L.; Hsu, C.C.; Paez, J.S.; Pan, L.; Andaluz, H.; Wendt, M.K.; Iliuk, A.B.; Zhu, J.K.; Tao, W.A. Phosphoproteins in Extracellular Vesicles as Candidate Markers for Breast Cancer. *Proc. Natl. Acad. Sci. USA* **2017**, *114*, 3175–3180. [CrossRef]
49. Manek, R.; Moghieb, A.; Yang, Z.; Kumar, D.; Kobessiy, F.; Sarkis, G.A.; Raghavan, V.; Wang, K.K.W. Protein Biomarkers and Neuroproteomics Characterization of Microvesicles/Exosomes from Human Cerebrospinal Fluid Following Traumatic Brain Injury. *Mol. Neurobiol.* **2018**, *55*, 6112–6128. [CrossRef]
50. Greening, D.W.; Xu, R.; Ji, H.; Tauro, B.J.; Simpson, R.J. A Protocol for Exosome Isolation and Characterization: Evaluation of Ultracentrifugation, Density-Gradient Separation, and Immunoaffinity Capture Methods. *Methods Mol. Biol.* **2015**, *1295*, 179–209.
51. Iwai, K.; Minamisawa, T.; Suga, K.; Yajima, Y.; Shiba, K. Isolation of Human Salivary Extracellular Vesicles by Iodixanol Density Gradient Ultracentrifugation and Their Characterizations. *J. Extracell. Vesicles* **2016**, *5*, 30829. [CrossRef] [PubMed]

52. Arab, T.; Raffo-Romero, A.; Van Camp, C.; Lemaire, Q.; Le Marrec-Croq, F.; Drago, F.; Aboulouard, S.; Slomianny, C.; Lacoste, A.S.; Guigon, I.; et al. Proteomic Characterisation of Leech Microglia Extracellular Vesicles (Evs): Comparison between Differential Ultracentrifugation and Optiprep (Tm) Density Gradient Isolation. *J. Extracell. Vesicles* **2019**, *8*, 1603048. [CrossRef] [PubMed]
53. Musante, L.; Saraswat, M.; Duriez, E.; Byrne, B.; Ravida, A.; Domon, B.; Holthofer, H. Biochemical and Physical Characterisation of Urinary Nanovesicles Following Chaps Treatment. *PLoS ONE* **2012**, *7*, e37279. [CrossRef] [PubMed]
54. Barrachina, M.N.; Sueiro, A.M.; Casas, V.; Izquierdo, I.; Hermida-Nogueira, L.; Guitian, E.; Casanueva, F.F.; Abian, J.; Carrascal, M.; Pardo, M.; et al. A Combination of Proteomic Approaches Identifies a Panel of Circulating Extracellular Vesicle Proteins Related to the Risk of Suffering Cardiovascular Disease in Obese Patients. *Proteomics* **2019**, *19*, e1800248. [CrossRef] [PubMed]
55. Lobb, R.J.; Becker, M.; Wen, S.W.; Wong, C.S.F.; Wiegmans, A.P.; Leimgruber, A.; Moller, A. Optimized Exosome Isolation Protocol for Cell Culture Supernatant and Human Plasma. *J. Extracell. Vesicles* **2015**, *4*, 27031. [CrossRef] [PubMed]
56. Merchant, M.L.; Powell, D.W.; Wilkey, D.W.; Cummins, T.D.; Deegens, J.K.; Rood, I.M.; McAfee, K.J.; Fleischer, C.; Klein, E.; Klein, J.B. Microfiltration Isolation of Human Urinary Exosomes for Characterization by MS. *Proteom. Clin. Appl.* **2010**, *4*, 84–96. [CrossRef]
57. Musante, L.; Tataruch, D.; Gu, D.F.; Benito-Martin, A.; Calzaferri, G.; Aherne, S.; Holthofer, H. A Simplified Method to Recover Urinary Vesicles for Clinical Applications, and Sample Banking. *Sci. Rep.* **2014**, *4*, 7532. [CrossRef]
58. Hu, S.; Musante, L.; Tataruch, D.; Xu, X.; Kretz, O.; Henry, M.; Meleady, P.; Luo, H.; Zou, H.; Jiang, Y.; et al. Purification and Identification of Membrane Proteins from Urinary Extracellular Vesicles Using Triton X-114 Phase Partitioning. *J. Proteome Res.* **2018**, *17*, 86–96. [CrossRef]
59. Heinemann, M.L.; Ilmer, M.; Silva, L.P.; Hawke, D.H.; Recio, A.; Vorontsova, M.A.; Alt, E.; Vykoukal, J. Benchtop Isolation and Characterization of Functional Exosomes by Sequential Filtration. *J. Chromatogr. A* **2014**, *1371*, 125–135. [CrossRef]
60. Osti, D.; Del Bene, M.; Rappa, G.; Santos, M.; Matafora, V.; Richichi, C.; Faletti, S.; Beznoussenko, G.V.; Mironov, A.; Bachi, A.; et al. Clinical Significance of Extracellular Vesicles in Plasma from Glioblastoma Patients. *Clin. Cancer Res.* **2019**, *25*, 266–276. [CrossRef]
61. Musante, L.; Tataruch-Weinert, D.; Kerjaschki, D.; Henry, M.; Meleady, P.; Holthofer, H. Residual Urinary Extracellular Vesicles in Ultracentrifugation Supernatants after Hydrostatic Filtration Dialysis Enrichment. *J. Extracell. Vesicles* **2017**, *6*, 1267896. [CrossRef] [PubMed]
62. De Menezes-Neto, A.; Saez, M.J.F.; Lozano-Ramos, I.; Segui-Barber, J.; Martin-Jaular, L.; Ullate, J.M.E.; Fernandez-Becerra, C.; Borras, F.E.; del Portillo, H.A. Size-Exclusion Chromatography as a Stand-Alone Methodology Identifies Novel Markers in Mass Spectrometry Analyses of Plasma-Derived Vesicles from Healthy Individuals. *J. Extracell. Vesicles* **2015**, *4*, 27378. [CrossRef] [PubMed]
63. Karimi, N.; Cvjetkovic, A.; Jang, S.C.; Crescitelli, R.; Hosseinpour Feizi, M.A.; Nieuwland, R.; Lotvall, J.; Lasser, C. Detailed Analysis of the Plasma Extracellular Vesicle Proteome after Separation from Lipoproteins. *Cell Mol. Life Sci.* **2018**, *75*, 2873–2886. [CrossRef] [PubMed]
64. Smolarz, M.; Pietrowska, M.; Matysiak, N.; Mielanczyk, L.; Widlak, P. Proteome Profiling of Exosomes Purified from a Small Amount of Human Serum: The Problem of Co-Purified Serum Components. *Proteomes* **2019**, *7*, 18. [CrossRef] [PubMed]
65. Aqrawi, L.A.; Galtung, H.K.; Vestad, B.; Ovstebo, R.; Thiede, B.; Rusthen, S.; Young, A.; Guerreiro, E.M.; Utheim, T.P.; Chen, X.; et al. Identification of Potential Saliva and Tear Biomarkers in Primary Sjogren's Syndrome, Utilising the Extraction of Extracellular Vesicles and Proteomics Analysis. *Arthritis Res. Ther.* **2017**, *19*, 14. [CrossRef]
66. Foers, A.D.; Chatfield, S.; Dagley, L.F.; Scicluna, B.J.; Webb, A.I.; Cheng, L.; Hill, A.F.; Wicks, I.P.; Pang, K.C. Enrichment of Extracellular Vesicles from Human Synovial Fluid Using Size Exclusion Chromatography. *J. Extracell. Vesicles* **2018**, *7*, 1490145. [CrossRef] [PubMed]
67. Chen, Y.Y.; Xie, Y.; Xu, L.; Zhan, S.H.; Xiao, Y.; Gao, Y.P.; Wu, B.; Ge, W. Protein Content and Functional Characteristics of Serum-Purified Exosomes from Patients with Colorectal Cancer Revealed by Quantitative Proteomics. *J. Extracell. Vesicles* **2017**, *140*, 900–913. [CrossRef]

68. Zhang, W.; Ou, X.; Wu, X. Proteomics Profiling of Plasma Exosomes in Epithelial Ovarian Cancer: A Potential Role in the Coagulation Cascade, Diagnosis and Prognosis. *Int. J. Oncol.* **2019**, *54*, 1719–1733. [CrossRef]
69. Tsuno, H.; Arito, M.; Suematsu, N.; Sato, T.; Hashimoto, A.; Matsui, T.; Omoteyama, K.; Sato, M.; Okamoto, K.; Tohma, S.; et al. A Proteomic Analysis of Serum-Derived Exosomes in Rheumatoid Arthritis. *BMC Rheumatol.* **2018**, *2*, 35. [CrossRef]
70. Leberman, R. The Isolation of Plant Viruses by Means of "Simple" Coacervates. *Virology* **1966**, *30*, 341–347. [CrossRef]
71. Ding, M.; Wang, C.; Lu, X.; Zhang, C.; Zhou, Z.; Chen, X.; Zhang, C.Y.; Zen, K.; Zhang, C. Comparison of Commercial Exosome Isolation Kits for Circulating Exosomal Microrna Profiling. *Anal. Bioanal. Chem.* **2018**, *410*, 3805–3814. [CrossRef] [PubMed]
72. Weng, Y.J.; Sui, Z.G.; Shan, Y.C.; Hu, Y.C.; Chen, Y.B.; Zhang, L.H.; Zhang, Y.K. Effective Isolation of Exosomes with Polyethylene Glycol from Cell Culture Supernatant for in-Depth Proteome Profiling. *Analyst* **2016**, *141*, 4640–4646. [CrossRef] [PubMed]
73. Rider, M.A.; Hurwitz, S.N.; Meckes, D.G. Extrapeg: A Polyethylene Glycol-Based Method for Enrichment of Extracellular Vesicles. *Sci. Rep.* **2016**, *6*, 23978. [CrossRef] [PubMed]
74. Shin, H.; Han, C.; Labuz, J.M.; Kim, J.; Kim, J.; Cho, S.; Gho, Y.S.; Takayama, S.; Park, J. High-Yield Isolation of Extracellular Vesicles Using Aqueous Two-Phase System. *Sci. Rep.* **2015**, *5*, 13103. [CrossRef] [PubMed]
75. Wang, D.; Sun, W. Urinary Extracellular Microvesicles: Isolation Methods and Prospects for Urinary Proteome. *Proteomics* **2014**, *14*, 1922–1932. [CrossRef] [PubMed]
76. Hildonen, S.; Skarpen, E.; Halvorsen, T.G.; Reubsaet, L. Isolation and Mass Spectrometry Analysis of Urinary Extraexosomal Proteins. *Sci. Rep.* **2016**, *6*, 36331. [CrossRef] [PubMed]
77. Ueda, K.; Ishikawa, N.; Tatsuguchi, A.; Saichi, N.; Fujii, R.; Nakagawa, H. Antibody-Coupled Monolithic Silica Microtips for Highthroughput Molecular Profiling of Circulating Exosomes. *Sci. Rep.* **2014**, *4*, 6232. [CrossRef] [PubMed]
78. Taylor, D.D.; Gercel-Taylor, C. Microrna Signatures of Tumor-Derived Exosomes as Diagnostic Biomarkers of Ovarian Cancer. *Gynecol. Oncol.* **2008**, *110*, 13–21. [CrossRef]
79. Kalra, H.; Adda, C.G.; Liem, M.; Ang, C.S.; Mechler, A.; Simpson, R.J.; Hulett, M.D.; Mathivanan, S. Comparative Proteomics Evaluation of Plasma Exosome Isolation Techniques and Assessment of the Stability of Exosomes in Normal Human Blood Plasma. *Proteomics* **2013**, *13*, 3354–3364. [CrossRef]
80. Tauro, B.J.; Greening, D.W.; Mathias, R.A.; Mathivanan, S.; Ji, H.; Simpson, R.J. Two Distinct Populations of Exosomes Are Released from Lim1863 Colon Carcinoma Cell-Derived Organoids. *Mol. Cell. Proteom.* **2013**, *12*, 587–598. [CrossRef]
81. Ghosh, A.; Davey, M.; Chute, I.C.; Griffiths, S.G.; Lewis, S.; Chacko, S.; Barnett, D.; Crapoulet, N.; Fournier, S.; Joy, A.; et al. Rapid Isolation of Extracellular Vesicles from Cell Culture and Biological Fluids Using a Synthetic Peptide with Specific Affinity for Heat Shock Proteins. *PLoS ONE* **2014**, *9*, e110443. [CrossRef] [PubMed]
82. Bijnsdorp, I.V.; Maxouri, O.; Kardar, A.; Schelfhorst, T.; Piersma, S.R.; Pham, T.V.; Vis, A.; van Moorselaar, R.J.; Jimenez, C.R. Feasibility of Urinary Extracellular Vesicle Proteome Profiling Using a Robust and Simple, Clinically Applicable Isolation Method. *J. Extracell. Vesicles* **2017**, *6*, 1313091. [CrossRef] [PubMed]
83. Balaj, L.; Atai, N.A.; Chen, W.L.; Mu, D.; Tannous, B.A.; Breakefield, X.O.; Skog, J.; Maguire, C.A. Heparin Affinity Purification of Extracellular Vesicles. *Sci. Rep.* **2015**, *5*, 10266. [CrossRef] [PubMed]
84. Gao, F.Y.; Jiao, F.L.; Xia, C.S.; Zhao, Y.; Ying, W.T.; Xie, Y.P.; Guan, X.Y.; Tao, M.; Zhang, Y.J.; Qin, W.J.; et al. A Novel Strategy for Facile Serum Exosome Isolation Based on Specific Interactions between Phospholipid Bilayers and TiO_2. *Chem. Sci.* **2019**, *10*, 1579–1588. [CrossRef] [PubMed]
85. Tan, K.H.; Tan, S.S.; Sze, S.K.; Lee, W.K.R.; Ng, M.J.; Lim, S.K. Plasma Biomarker Discovery in Preeclampsia Using a Novel Differential Isolation Technology for Circulating Extracellular Vesicles. *Am. J. Obstet. Gynecol.* **2014**, *211*, 380-e1. [PubMed]
86. Nakai, W.; Yoshida, T.; Diez, D.; Miyatake, Y.; Nishibu, T.; Imawaka, N.; Naruse, K.; Sadamura, Y.; Hanayama, R. A Novel Affinity-Based Method for the Isolation of Highly Purified Extracellular Vesicles. *Sci. Rep.* **2016**, *6*, 33935. [CrossRef] [PubMed]
87. Takov, K.; Yellon, D.M.; Davidson, S.M. Comparison of Small Extracellular Vesicles Isolated from Plasma by Ultracentrifugation or Size-Exclusion Chromatography: Yield, Purity and Functional Potential. *J. Extracell. Vesicles* **2019**, *8*, 1560809. [CrossRef] [PubMed]

88. Davis, C.N.; Phillips, H.; Tomes, J.J.; Swain, M.T.; Wilkinson, T.J.; Brophy, P.M.; Morphew, R.M. The Importance of Extracellular Vesicle Purification for Downstream Analysis: A Comparison of Differential Centrifugation and Size Exclusion Chromatography for Helminth Pathogens. *PLoS Negl. Trop. Dis.* **2019**, *13*, e0007191. [CrossRef]
89. Stranska, R.; Gysbrechts, L.; Wouters, J.; Vermeersch, P.; Bloch, K.; Dierickx, D.; Andrei, G.; Snoeck, R. Comparison of Membrane Affinity-Based Method with Size-Exclusion Chromatography for Isolation of Exosome-Like Vesicles from Human Plasma. *J. Transl. Med.* **2018**, *16*, 1. [CrossRef]
90. Patel, G.K.; Khan, M.A.; Zubair, H.; Srivastava, S.K.; Khushman, M.; Singh, S.; Singh, A.P. Comparative Analysis of Exosome Isolation Methods Using Culture Supernatant for Optimum Yield, Purity and Downstream Applications. *Sci. Rep.* **2019**, *9*, 5335. [CrossRef]
91. Antounians, L.; Tzanetakis, A.; Pellerito, O.; Catania, V.D.; Sulistyo, A.; Montalva, L.; McVey, M.J.; Zani, A. The Regenerative Potential of Amniotic Fluid Stem Cell Extracellular Vesicles: Lessons Learned by Comparing Different Isolation Techniques. *Sci. Rep.* **2019**, *9*, 1837. [CrossRef] [PubMed]
92. Royo, F.; Zuniga-Garcia, P.; Sanchez-Mosquera, P.; Egia, A.; Perez, A.; Loizaga, A.; Arceo, R.; Lacasa, I.; Rabade, A.; Arrieta, E.; et al. Different Ev Enrichment Methods Suitable for Clinical Settings Yield Different Subpopulations of Urinary Extracellular Vesicles from Human Samples. *J. Extracell. Vesicles* **2016**, *5*, 29497. [CrossRef] [PubMed]
93. Andreu, Z.; Rivas, E.; Sanguino-Pascual, A.; Lamana, A.; Marazuela, M.; Gonzalez-Alvaro, I.; Sanchez-Madrid, F.; de la Fuente, H.; Yanez-Mo, M. Comparative Analysis of Ev Isolation Procedures for Mirnas Detection in Serum Samples. *J. Extracell. Vesicles* **2016**, *5*, 31655. [CrossRef] [PubMed]
94. Rood, I.M.; Deegens, J.K.; Merchant, M.L.; Tamboer, W.P.; Wilkey, D.W.; Wetzels, J.F.; Klein, J.B. Comparison of Three Methods for Isolation of Urinary Microvesicles to Identify Biomarkers of Nephrotic Syndrome. *Kidney Int.* **2010**, *78*, 810–816. [CrossRef] [PubMed]
95. Caradec, J.; Kharmate, G.; Hosseini-Beheshti, E.; Adomat, H.; Gleave, M.; Guns, E. Reproducibility and Efficiency of Serum-Derived Exosome Extraction Methods. *Clin. Biochem.* **2014**, *47*, 1286–1292. [CrossRef] [PubMed]
96. Joy, A.P.; Ayre, D.C.; Chute, I.C.; Beauregard, A.P.; Wajnberg, G.; Ghosh, A.; Lewis, S.M.; Ouellette, R.J.; Barnett, D.A. Proteome Profiling of Extracellular Vesicles Captured with the Affinity Peptide Vn96: Comparison of Laemmli and Trizol (C) Protein-Extraction Methods. *J. Extracell. Vesicles* **2018**, *7*, 1438727. [CrossRef] [PubMed]
97. Fel, A.; Lewandowska, A.E.; Petrides, P.E.; Wisniewski, J.R. Comparison of Proteome Composition of Serum Enriched in Extracellular Vesicles Isolated from Polycythemia Vera Patients and Healthy Controls. *Proteomes* **2019**, *7*, 20. [CrossRef]
98. Xie, X.F.; Chu, H.J.; Xu, Y.F.; Hua, L.; Wang, Z.P.; Huang, P.; Jia, H.L.; Zhang, L. Proteomics Study of Serum Exosomes in Kawasaki Disease Patients with Coronary Artery Aneurysms. *Cardiol. J.* **2018**. [CrossRef]
99. Gonzales, P.A.; Pisitkun, T.; Hoffert, J.D.; Tchapyjnikov, D.; Star, R.A.; Kleta, R.; Wang, N.S.; Knepper, M.A. Large-Scale Proteomics and Phosphoproteomics of Urinary Exosomes. *J. Am. Soc. Nephrol.* **2009**, *20*, 363–379. [CrossRef]
100. Kittivorapart, J.; Crew, V.K.; Wilson, M.C.; Heesom, K.J.; Siritanaratkul, N.; Toye, A.M. Quantitative Proteomics of Plasma Vesicles Identify Novel Biomarkers for Hemoglobin E/Beta-Thalassemic Patients. *Blood Adv.* **2018**, *2*, 95–104. [CrossRef]
101. Sok Hwee Cheow, E.; Hwan Sim, K.; de Kleijn, D.; Neng Lee, C.; Sorokin, V.; Sze, S.K. Simultaneous Enrichment of Plasma Soluble and Extracellular Vesicular Glycoproteins Using Prolonged Ultracentrifugation-Electrostatic Repulsion-Hydrophilic Interaction Chromatography (Puc-Erlic) Approach. *Mol. Cell. Proteom.* **2015**, *14*, 1657–1671. [CrossRef] [PubMed]
102. Saraswat, M.; Joenvaara, S.; Musante, L.; Peltoniemi, H.; Holthofer, H.; Renkonen, R. N-Linked (N-) Glycoproteomics of Urinary Exosomes. *Mol. Cell. Proteom.* **2015**, *14*, 263–276. [CrossRef] [PubMed]
103. Braga-Lagache, S.; Buchs, N.; Iacovache, M.I.; Zuber, B.; Jackson, C.B.; Heller, M. Robust Label-Free, Quantitative Profiling of Circulating Plasma Microparticle (Mp) Associated Proteins. *Mol. Cell. Proteom.* **2016**, *15*, 3640–3652. [CrossRef] [PubMed]
104. Kodidela, S.; Wang, Y.; Patters, B.J.; Gong, Y.; Sinha, N.; Ranjit, S.; Gerth, K.; Haque, S.; Cory, T.; McArthur, C.; et al. Proteomic Profiling of Exosomes Derived from Plasma of Hiv-Infected Alcohol Drinkers and Cigarette Smokers. *J. Neuroimmune Pharmacol.* **2019**. [CrossRef] [PubMed]

105. Wang, T.T.; Anderson, K.W.; Turko, I.V. Assessment of Extracellular Vesicles Purity Using Proteomic Standards. *Analyt. Chem.* **2017**, *89*, 11070–11075. [CrossRef]
106. Wang, T.; Turko, I.V. Proteomic Toolbox to Standardize the Separation of Extracellular Vesicles and Lipoprotein Particles. *J. Proteome Res.* **2018**, *17*, 3104–3113. [CrossRef]

MDPI
St. Alban-Anlage 66
4052 Basel
Switzerland
Tel. +41 61 683 77 34
Fax +41 61 302 89 18
www.mdpi.com

Molecules Editorial Office
E-mail: molecules@mdpi.com
www.mdpi.com/journal/molecules

www.ingramcontent.com/pod-product-compliance
Lightning Source LLC
LaVergne TN
LVHW070839170726
843515LV00005B/526

* 9 7 8 3 0 3 9 3 6 5 7 7 7 *